Professor Samuel F. B. Morse, c. 1850. Carte de visite, published by E. Anthony, NY, from a Matthew B. Brady negative.

LOVE LETTERS TO SPIKE
A Telegrapher's Lament

With a Brief, Eclectic
History
of
Communications
in the Seacoast

By
Bill Holly
K1BH

Placenames Press
Portsmouth, New Hampshire

LOVE LETTERS TO SPIKE: A Telegrapher's Lament

ISBN 0-9767590-5-5

PLACENAMES PRESS
170 Mechanic Street
Portsmouth, NH 03801
www.placenamespress.com

Printed in the United States of America

Cover and book design by Nancy Grossman

Library of Congress Control Number 2005910820

Cover image: USS *Faraday*. Period cabinet photograph c1875-1880. Photographer unidentified, possibly Davis Bros.

Table of Contents

Illustrations

Acknowledgements

I wish to thank Tom Hardiman and his staff at the Portsmouth Athenaeum for their assistance and good advice. Wayne Manson of the Kittery Historical and Naval Museum and Walter Ross and Jim Dolph of the Portsmouth Naval Shipyard Museum and Visitor Center contributed material assistance to this project. Thanks also to Carol Bovee, archivist at the First United Methodist Church of Portsmouth. The staffs of the Portsmouth Public Library; Rice Public Library, Kittery, ME; the New Bedford Public Library, New Bedford, MA; the Connecticut State Library, Hartford; and the clerks at Portsmouth, NH, city hall all made important contributions. I have to thank Joseph Frost of Eliot, ME, historian and role model, for his guidance and material on Moses Farmer. Finally, none of this could have been accomplished without the support and assistance of my friend and companion for the last forty years, Fran Holly.

Photos unless specifically attributed were either taken by the author or are from his collection of post cards, stereo views, cartes de visite and cabinet photographs.

Introduction

Depending on the historian telling the story, electrical communications was either the chicken that led the charge in the industrial revolution or simply one of the many eggs spawned in that upheaval. Whichever view you favor, it would be hard to comprehend how today's world could operate without the progeny of those simple electro-magnetic telegraph systems that blossomed on both sides of the Atlantic in the mid 1830s and 40s. The telephone, radio, TV, computers and their networks all trace their lineage directly back to those embryonic telegraphs.

While at the Portsmouth Athenaeum researching the role the New Hampshire Seacoast played in the history of electric communications, I discovered a packet of letters written by a young telegrapher at the Portsmouth Navy Yard to his fiancée in Hartford, CT, during the spring of 1914. These letters contain a running account of the telegrapher's day to day activities at work and play. They are a window into the life of an average lower middle class working man in pre-World War I Portsmouth, NH. These letters also tell us something of the telegrapher's profession, a craft that no longer exists but was a prerequisite to all of the electrical communications disciplines which exist today.

Finally, in his daily activities our telegrapher often touched on the rich history of communications in the Seacoast. First, simply being a Morse operator brings to mind that Samuel F.B. Morse, father of the American telegraph, began his first career, as an artist, in New Hampshire. Morse spent several weeks in 1816-17 in Portsmouth painting portraits and later married a Concord, NH, woman. Speaking of women, our telegrapher mentions in one of his letters meeting an ex lady telegrapher. As it turns out, one of the first female telegraphers in the country learned her craft in Dover, NH, and managed the Portsmouth office for several years in the 1855-1865 period.

Our telegrapher visited the cable station in Rye, NH, which served the fourth of the the great trans-Atlantic cables to be laid between Europe and America. He also befriended the wireless operators at the Navy Yard wireless station NAC which was part of the U.S. Navy's first network of radio stations used to maintain communications between Navy vessels at sea and Naval headquarters in Washington. This was during the infancy of a technology that now allows us to routinely talk and send pictures around the world and even to other planets.

So, let's meet this young telegrapher and his fiancée, snoop through his correspondence, and examine what it tells us about 1914 Portsmouth and the history of New England electronic communications in general.

Chapter 1

Herbert Denison Waldron was born on April 8, 1888, in Hartford, CT. He was the only child of Arthur M. Waldron and Abbie Anna Denison Waldron. Arthur was an operator for the Western Union Telegraph Co. and worked his way up to Division Director before he retired. Herbert became an operator when he was twenty years old and in the 1909 Hartford directory was listed as an operator at the same office in which his father was an Assistant Chief Operator of Western Union. By the time Herbert arrived in Portsmouth, in March of 1914, he had five years experience as a telegrapher. Waldron's letters never reveal exactly why he quit Western Union and came to Portsmouth but the tone of the letters hint broadly that this was not a career change of his choosing.

Herbert Waldron's fiancée, Grace I. Glen, was born in 1885 at Hartford, CT, the younger daughter of Scots immigrant Agnes Fleming Glen and Irish-born Samuel Glen, house painter. Grace was brought up in an extended family which included her grandparents Charles and Christine Fleming; her uncle, Agnes' brother, Thomas Fleming; her mother and father, and her sister, Christina F. Glen, who was 15 years older than Grace. Like her parents Christina was born in the old country. Grace worked in her uncle's grocery store as a bookkeeper when she was in her early teens.

Waldron's letters give no clue as to how he and Grace met, but the Hartford city directories of the late 1800s indicate that the Waldron and Fleming/Glen families lived within two city blocks of each other for a while. It also appears that the Fleming/Glens had a male boarder for a time who was a telegrapher. Either one of these circumstances could have precipitated an event that led to Grace and Herbert's introduction but it's unlikely we shall ever know for sure. What we do know is that on March 16th, a little less than a month shy of his 26th birthday, Herbert Waldron reported for his first day of work as a telegrapher in the Commandant's office of the Portsmouth Navy Yard. On March 17th he used the office typewriter to write the first of 34 letters to Grace. About half of Herb's letters were typed on the office typewriter while he was at work. Let's read the letters and find out what happens to the young couple and see what these letters reveal about Portsmouth and its communications history.

Note: Each of Waldron's letters has been carefully transcribed. All typographical errors in the text are faithful to the original. All parentheses with the letters are Waldron's own. Any items in brackets have been added by the author for clarity's sake. Likewise, this applies to any other original handwritten material and newspaper articles of the day.

Chapter 2

Portsmouth N H
Mar. 17, 1914

Dearest Gracie,

Have not got quite settled yet and so have no permanent address, except the Nave Yard. This letter is a sad one as I am so homesick that I am unable to write you a decent letter. I have lost about five pounds since I arrived. The job is quite an easy one but they tell me it is hard to gwt a raise in pay, as the Navy Yard is the poorest paying branch of the service.

I know you will not like it here as none of the fellows wives seem to be able to stand it. Al and I met her friend in Boston at the depot and she says she is so homesick that she positively cannot stand Boston another year and her hubby is looking for work in Groton. There is a government ferry that runs over to the island leaves here at 845 am and no charges, returning at 440 pm, making my hours from 9 am till 430 pm with a half hour for lunch at twelve.

There is a lunch room operated here where they serve mostly beans, in a camp-like style. If my attitude does not change toward this place I will beat it just as soon as I get back on my feet and a new suit of clothes. One of the fellows was telling me that I could buy a good suit here for $9. The town is only about the same size as New Britain [Connecticut, ed.] in appearance through the business section, but has six hotels and three moving picture houses. I am working in the city hall of Kittery Island and it resembles our own city hall in appearance and age, but we have fine ventilation and can smoke our heads off with our feet posted up on the radiator. Pretty soft cutting out the home-sickness part but it looks like the same money for several years. Doesnt cost much to live here as they only have three things, morning, noon, and night. I am staying on a chicken farm now and was awakened at three am this morning by the crowing of the roosters. Not a thing to go to. But they have a fine new methodist church here and Mr. Bodwell is a strong methodist and church worker, married with one child.

The way I now feel this will be about all, hoping to hear from you asking me to return home, with lots of love to you and regards to the whole family, I remain,

Your Herbert

Building 13, where Herbert Waldron worked. Postcard view c. 1914.

Waldron is right about low wages. He was working a seven-hour day; at three dollars a day, that's only forty two cents an hour. Typical Western Union and Railroad Telegraphers of this period earned fifty to sixty cents an hour.

Waldron's landlord, Clarence P. Bodwell, occupied the left side of a side-by-side two-family home at 48 Orchard St. Our telegrapher's description of Portsmouth appears fairly accurate. There were seven hotels advertised in the Portsmouth city directory for 1914; however, there were only two movie houses in the directory, the Portsmouth Theatre, on Chestnut near Congress (today The Music Hall), and the Premier Scenic Temple at 28 High Street. Waldron mentions the Pierce Hall (presumably in the Pierce Building adjacent to the Athenaeum in Market Square) also showing movies.

This was the period when the movies were just emerging as a force in the entertainment field. There were no "movie theaters" as such. Two and three-reeler silent films of ten to twenty minutes duration were being shown in store fronts, minstrel and so called "legitimate" theaters. The first full-length feature film, D.W. Griffith's "Birth of a Nation," was a year away, as were the first theaters built expressly to show motion pictures.

48-50 Orchard Street, Portsmouth, New Hampshire, where Waldron lived through the spring of 1914, as it appears today.

The United Methodist Church on Miller Avenue, Portsmouth, NH, was built in 1912, just two years before Waldron arrived in town. Formerly the Methodist Episcopal Church.

The Music Hall on Chestnut Street as it looked in May, 2004. In 1914, it was the Portsmouth Theater, home to vaudeville and one and two-reeler motion pictures.

Chapter 3

Portsmouth N H
Mar 17th

Dear Gracie,

The first thing please pardon this use of the paper as I came away from the factory in a hurry and forgot to bring some home. I am staying at Bodwells (methodist and high mason) for this week which is located one mile from dock and rather bad walking too. I work very hard trying to kill time but aside from telegraphing answering phone calls on my own desk, I assume the duty of filing letters. Now this last task is not difficult but the hard part is to file them under their proper heads picking out and classifying their subjects. I should worry. We have two ships docked for repairs, Tacoma; Brutus both of which have telephones. When I get a phone I must ans. it "Commandants office," pronounced not like spelled but Ccomma dants" or better "Commer dants." Je! Gosh thats tuff stuff. The town proper consists: - 4 W.U. Tel. ofs's; [Western Union Telegraph Offices] 1 Postal [Postal Telegraph]; 6 hotels (or apologys); 3 Movies; 6 Resturants; Foggy weather; 1 River; 4 trolley lines; 200 Drug stores; 1000 saloons; 2 breweries; some micks; plenty of air; few sidewalks; and a Navy yard. Je Spike I cant live without seeing you and you couldnt live without Hfd. [Hartford] Herewith I submit a rough sketch [not in the file of letters] of our famous town showing prin. streets and route to my home. I leave on the boat at 845 am & stay till 440 Pm. The Island is only about 1 mile from the town and is 2 miles square. Very swift current in river. Tide drops 20 feet. With love from yours,

Herbert

The telephone was well entrenched on the yard by this time. An entry from a yard log at the Kittery Naval & Historical Society Museum states: "1882...First telephone in use connecting Yard to Portsmouth."

Waldron mentions two ships being in the Yard when he started work. The *Brutus*, a two thousand ton, 332' 6" cargo vessel, was being used to supply vessels of the Atlantic Fleet that were operating in Mexican waters. The *Tacoma*, a thirty-two hundred ton 308' cruiser, had come to the Navy Yard in January for repairs and in May resumed operations in Mexican waters.

USS Tacoma, *cruiser 3rd class. Built in San Francisco, 1903. Postcard view, c. 1910.*

The two hundred drug stores and one thousand bars Waldron mentions are hyperbole, but he undercounted the breweries and hotels. The 1914 Portsmouth City Directory lists five brewing companies and seven hotels. There were four telegraph offices in the city, Western Union offices at 29 Congress St. and 22 Daniel St., a Postal Telegraph & Cable Co. at 10 Congress, and the Boston and Maine railroad station also had a branch Western Union office. Herb's use of the word "mick" (a derogatory term for Irishmen) probably only marks him as a product of his time. Derogatory descriptors for every minority group were so commonly used that they appeared regularly in almost every print format from advertising broadsides to newspapers, to fiction and non-fiction books. It does make one wonder, given her Scots-Irish heritage how well that comment went over with Grace!

Unfortunately Waldron's map of Portsmouth does not survive. His estimate of the tides is also a bit inflated; eight to ten feet would be closer to the mark. There was only one trolley line, the Portsmouth Electric

Railway, but it had four routes: the Main Line from Portsmouth to Rye Center, Rye Beach and Portsmouth Junction (North Beach), 10.33 miles; the North Hampton branch, 2.62 miles; the Portsmouth Plains, 4.99 miles; and the Christian Shore, 2.56 miles, for a total of 18.10 miles. The Portsmouth Electric Railway, built and maintained by the Boston and Maine Railroad, was never profitable. It did well in the summer, running many special excursion cars to the beaches on the weekends. The winter ridership, however, was very light. Portsmouth was a small city of 12,000 people, was compactly built and most destinations were within a reasonable walking distance. (A one or two mile walk was not considered an extreme hardship in 1914!) The introduction of Ford's Model T in 1913 was already creating inroads on the trolley business in 1914. Riders and service declined in lockstep. The Portsmouth Electric Railway lasted another decade before the Boston & Maine pulled the plug for good in 1925.

Trolley wires dominate the sky in downtown Portsmouth. Postcard view, c. 1910.

Interestingly, the very first electric railway car was invented and first exhibited in Dover, New Hampshire, in 1847. Moses Garish Farmer was born in Boscawen, NH, February 9, 1820. He attended school at Boscawen Academy, Phillips Academy, and Dartmouth, where ill health forced him to drop out before graduation. Farmer taught in Eliot, ME, then Dover, NH, while pursuing his own studies in mathematics and electricity.

By 1847, Farmer had given up teaching to devote his full time to developing and promoting inventions in the electrical field. His first such invention was the electric railway car. Farmer was brilliant and soon applied his fertile mind to the newly-opened field of electro-magnetic telegraphy. He made numerous improvements to telegraphic instruments, supervised the building of telegraph lines from Boston to Vermont, and supervised the operation of some of these lines. He invented and built the first municipal fire alarm telegraph system. In April 1857, that system was demonstrated at Boston, MA. The United States Coast Survey selected Farmer to build a line connecting the Cambridge observatory with their Boston office. Also in the 1850s he built an electric clock and invented and successfully demonstrated duplex and quadruplex telegraph machines, which allowed two or four messages to be transmitted simultaneously over a single wire. He also worked on electrical lighting twenty years before Edison, developed electro-plating systems, household burglar alarms, and the electrical smelting of aluminum.

In 1869, Farmer was hired to inspect the Atlantic Telegraph Cable system from Newfoundland to New York and report on its condition and make recommendations for improvements. In 1872, he went to work for the U.S. Navy at Newport, RI, instructing officers in electrical science and the related chemistry. He also invented electrical devices for use with torpedoes.

Moses Garish Farmer was an electrical genius and recognized as such by his peers, including Edison and Sir William Thompson (Lord Kelvin). He died in Chicago on May 25, 1893, while preparing an exhibit of his inventions for the Columbian Exposition.

The first electrically operated rail car, a passenger car, invented by Moses G. Farmer. Photo from the collections of the Portsmouth Athenaeum, c. 1847.

Moses G. Farmer.

Moses O. Farmer, from The Rich Legacy: The Memories of Hannah Tobey Farmer.

Chapter 4

Portsmouth N H.
March 18, 1914

Dear Gracie,

It is now eleven oclock and snowing hard so I thought I might improve some of my spare time writing you a few lines. It is a shame to take money for this kind of work but it certainly is too soft. I have considerable red tape work to do but can stall on it making believe I am busy on the wires. I wrote you last night and mailed it this morning but guess you will not receive it till this evening or perhaps tomorrow morning. For the past two nights we have had a vary heavy fog, in fact it was so thick that one could scarcely see your hand before your face.

As I said in my last letter, I like my room out to Bodwells house right in the residential section but Oh such a walk. I am eating way down town in restaurants. You can secure eats at a regular boarding house for $3.50 per week and get rooms most any where down street. They tell me a married person can board and room out much more cheaper than to keep house at a rent of twenty dollars a month. The chief clerk says he prefers to room out with his wife eating at one of these boarding house than to pay rent coal bills and electricity.

In this respect it is much cheaper living here than in Hartford. One of the captains formerly here tried this stunt all the time here, but later was transferred south. I was looking at room in the center of the city for three dollars a week at Dr. Chases house, Osteopath, (Mrs.) but she is in an old red brick house very antique and have not decided to take it as yet. she showed me Capts. former room, which is large enough for five persons, some room. I cannot understand how his wife would be satisfied to live out like that with practically no one for a companion during the day or any house work. Je, thats a tuff life all right.

The chief clerk here, Mr. young, is my boss and there are not many catholics in our office, in fact I know of but one. He sits right beside me.

I hope you can persuade me to return to Hartford in your next letter, now that you have the history of this place and have already come to the conclusion that you would not live here.

Lots of love and kisses till I get back to the old town.

Herbert.

P. S. Please excuse typewritten letters
They are much easier to read than my hand
writing and quicker to write

Fifty years after these letters were written, I was a seagoing telegrapher and used to, in the quiet hours of the night watches, roll the log around the platen on my typewriter, slip in a blank sheet, and dash off letters to my fiancée. I used the same lame excuse almost verbatim. My wife always maintained that she was the only woman ever to receive typewritten love letters. Needless to say when I came across Herb's letters I couldn't wait to show her she was wrong. Wives are never wrong!

Chapter 5

Portsmouth N.H. Mar. 19, 1914

Dear Gracie,

I am writing you a few lines once more, and as yet have not heard a word from you. Have just been out exploring the Island and find many buildings of interest, probably fifteen in number. We have a prison, hospital, emergency hospital, reading room, pool room machine shops restaurant, prison boat, electric light plant, and several other buildings that I know nothing about. Several of the prisoners were out on the bank of the river throwing stone down the bank, and they are made to work hard every day, under guard of an officer. The wind is blowing about sixty miles an hour and the officer is dancing up and down to keep his blood in circulation.

I have been in all day and have had nothing to do but file a few letters away and eat my dinner. Had some roast lamb dressing and potatoes for lunch. We have two ships in dry dock at the present time. One leaves today. No-body in our office works for a living but they think they do. But it will be busy when we get some ships in port. Our telephone exchange reaches all over the yard connecting each ship and building. I dont feel very happy today as I am homesick and miss you too much.

Talk about mud. You dont know what mud is in Hartford. It is nine inches thick here and I guess I have ruined my shoes. Isnt that awful.

After looking up the record of some the operators in the past I find that Dwyer left and went back on the Railroad in his home town because there was more money in it and more chance of advancement. The commanding officer remarked this afternoon that he hoped I would stay as there had been so much changing around, and it was hard to find anyone willing to stay very long at a time. Walker preceeded Dwyer, and stayed perhaps the longest of anyone, three years off and on. He came from Texas and brought his wife with him, but owing to sickness in his family after being here two years took a leave of absence, and returned for a short period, then went back home to stay after his father died. Page either came before or after him. A nice young man, wire chief of the Western Union from Boston Mass. and he stayed only four or six months returning back into the telegraph service. The only reasons I can see are that the town is poor and unattractive, and the job is too soft. Walker studied short hand right in this office and got a good job eventually with the Texas Oil People.

Please write soon. My address is care of General Delivery for the present. Best regards to the folks and love to you. Much of it.

Herbert.

Ptsmth N.H.
Mar. 22, 1914

Dear Gracie,

Your letter recd. yesty. mng., telling about wedding. It was, mailed at 11 pm and I got it at 8am next day. I guess about the best time to send mail is at night as we will receive it next am.

I went to the movies and dance last night. This combo. all takes place in one hall, the first floor for dancing and the second contains seats for the show. It was a pretty good bill and if one gets tired looking at the picture you can watch 'em dance, both going on at the same time. They have a crack orchestra, which plays constantly, with practically no intermission, playing all the latest dance music, and some piano player, believe me he can ragtime anything. Two hours & half solid pictures, dance and such music, It cannot be classed as a classical programme but it has the time and snap that make you stamp your feet.There are only four players, piano, violin, cornet, and drum-xzylophone, but they make enough noise to count as double their size. How can they work so hard. It cant be did in Hartford.

We have four wireless oprs here, enlisted men, who seem to be pretty good sort of chaps, personally and professionally. I met three of 'em, such as Baker, Caldwell, and Ferson. Ferson rather appeals to me more than the other two and I am going over to the Island & have supper with 'em Monday night at 5 pm. These fellows are all enlisted men and live right near the station having their own cook etc. Ferson wants to take the Civil Service exam and get a job like mine. He is a married man but I don't know where his wife is. She couldn't live with him on the island. Pretty soft for him, eh! He considers his time spent in the Navy as wasted.

The lowest temp. here this winter was 24 below. Not much swimming here as the water is too cold even in summer. Very few people know how to swim. Its snowing again. My love and lots of kisses.

Herbert

Eleven years before this letter was written and just seven years after Guglielmo Marconi transmitted the first radio signal across Salisbury Plain, England, in 1896, an entry in the Daily Log Book, U.S. Navy Yard, Portsmouth, NH, reads: "Aug 24, 1903 First section of wireless staff raised on Seavey's Id. near Reservoir." At the same time, the wireless station, building #103, was also being constructed. Wireless equipment from the German firm, Slaby-Arco, was installed. The station was assigned the call letters NAC, and became a part of the U.S. Navy's first wireless network along the East Coast. This network was designed to maintain communications with ships of the North Atlantic fleet and relay those messages back and forth to Washington.

The small white building to the left of the antenna mast and in front of the water tower was Building 103, the wireless building. Postcard view, c. 1910.

The original apparatus was unable to cover the distance between Portsmouth and Boston, though Slaby-Arco equipment could cover much greater distances from shore to ship; it was thought that the signals were meeting with interference caused by the land mass between Portsmouth and Boston. The Navy proposed to correct the problem by erecting a relay station at Thatcher's Island, the halfway point between the two cities.

The Cambridge, MA, firm Stone Telegraph & Telephone Co. made another proposal: they would install new equipment at Boston, Portsmouth, and Cape Cod at their own expense and give the Navy three months to test and see if it met the Navy's requirements without requiring an additional relay station at Thatcher's Island. This installation was completed on May 31, 1905. The equipment performed as advertised and was duly accepted by the Navy.

Building 103 was of wood construction fourteen by eighteen feet, eight foot to the eaves, and built on a stone foundation. The antenna structure was a standard Navy three-section mast, of the ship type, 180 feet tall.

Building 103's instruments were incredibly primitive. The transmitter consisted primarily of a high voltage transformer which, when turned on, impressed a voltage across two contacts so closely spaced that a spark would bridge the contacts as long as the power was on. By turning the transformer on and off in a pattern of Morse code dots and dashes, and applying the resulting pulses to an antenna, a crude signal is radiated. If the power applied is large, in this case seven thousand five hundred watts, the signal is radiated over a considerable distance.

Take a portable radio near a source of arcing electricity, an arc welder for instance, and listen to the hash on your radio. Now imagine that hash turned on and off in a series of dots and dashes. That was essentially what a transmitter was in 1905.

Receivers were equally primitive, comprised as they were of an antenna to receive the signal and one or two coils of wire wound on tubes that were suppose to "tune" the signal and separate it from others not wanted (usually unsuccessfully). The signal was then sent to a device to change the electrical energy from a radio frequency, inaudible to human ears, to an audio frequency and then to the ear phones.

The wireless industry was still in its infancy and was in a state of perpetual chaos. Many improvements were being introduced but, while one company held a significant patent for a receiver, another would hold the patents for a good transmitter. Since neither would cross-license to the other, it was not possible to take advantage of the best technology as it emerged. (Sounds surprisingly like the digital electronics business today.) Not so strangely, amateur radio operators often had better equipment than either commercial or military operators. Since amateurs mostly built their own equipment, they could employ the latest and best technology as soon as it was introduced, without regard as to who had a patent on what. This

Quarters 68 as it appears today. In 1914, it was home to the wireless operators, and is where Ferson and his mates entertained Waldron.

was the state of the electronics art in 1905, and it had not changed significantly by 1914.

When Ferson and his mates entertained Herb at Quarters 68 and then took him to Building 103 to listen in on the circuits there had been few changes. The original 180' mast had been replaced by two 152' masts separated by about two hundred feet and arranged so that a feed line attached to the wire between the two masts dropped directly down to the wireless building. There had been incremental improvements in equipment, but nothing of significance. The vacuum tube, the next quantum leap in the electronics time line, had been invented a decade earlier but still was in the experimental state. Well-to-do amateur radiomen and some forward-thinking commercial concerns were making limited use of vacuum tubes, but the technology was not in everyday use in the “Fleet”.

Chapter 6

Portsmouth N.H.,
March 23, 14.

Dear Grace,

I am going to stay at Bodwells house and they have agreed to furnish me with breakfast and suppers for this week on trial to see whether it agrees with me. I like the room very much although the bed is not up to the standard but fairly passable. It is about a mile walk to the dock and we have electric cars that run every half hour but it is no use to us to try to ride down as you never know when one is due and you can walk much quicker if you just miss one. So I will run up an enormous shoe bill to make up for the money saved on fares. It is lonesome up there as I have no young fellow chums, and I know the wireless men would like to have me board with them if the commandant would consent, but it is against the rules of the yard to have anyone there over night except enlisted men. They tried to do this same stunt with Dwyer as it would help out on their expenses and give them more and better quality of food, as their expense money is rather low after allowing for the cook.

My work consists of a lot of red tape filing letters and indexing which I dislike very much but there is no telegraphing to do so might as well be satisfied with something to take up my time. I just learned from good authority that no one enjoys bathing up here in the summer as the water is so cold so guess we will have to go to Marthas Vineyard for our annual bath and can only look upon this water. Business is dull today as we have only a couple ships in port for repairs but expect some before long.

Well I still have my same old suit on and it is holding out wonderfully and I expect it to last at least till I get paid the first , then I will do away with it entirely except the coat which I will use for office purposes. I will have to go to Boston for my clothes to get good ones at a moderate price. That is the way most of the fellows do that are wished on this place. I feel fairly well but not quite as well as when I am home, presume on account of eating around at the restaurant and I find most everybody complains the same way that do likewise. It keeps your stomach on the jump all the time, and when you enter the dining room your appetite vanishes out the back door or somewhere else. I have been around the accounting department and was introduced to most all the fellows, but none of them appeal especially to me and they are not over social among themselves, with the exception of the wireless men who are quite agreeable. I can get my board as reasonable here as any place I ever worked. Spent a mighty lonesome Sunday as it snowed all day. Remember me to the folks and Grandma. Love and kisses from your lonesome spike,

Herb.

Portsmouth N. H.
March 24, 1914

Dear Gracie,

Regret to say that I cannot return the piece about Chas. wedding which you ask for in your last letter, as I have either destroyed it or mislaid it. You should not have sent it without requesting the return of same in your first letter. Tell auntie that I have neglected her by not sending her a card, but never-the-less I often think of her, and how nice it would seem to run in and sample some of her cooking, and speak to you all. I will try and remember to mail her a card tonight or tomorrow.

Well last night I went up to the radio station for supper, and there were four of us fellows there, Webb of the Tacoma; Ferson, Baker and myself and we had some nice steak and beans apple pie and cheese and believe me it was some cooking. Mrs. Walker did it. She is the house-keeper for them. I listened in on the receiving phones at the station before and after supper. We also had a little poker game. Pretty soft for those men up there. Nothing but camp life with first class cooking. They all look healthy and have a good color. They are located right side of the prison, if you refer to Uncles card you can get an idea of the location.

The captain and chief clerk are trying to puzzle out, why it is that it is impossible to keep a telegrapher and clerk on the job more than a few weeks at a time. My job is the only one on the yard that the appointees do not die at the post. All clerkships that are filled never or rarely ever become vacant as they die on the job. That one job is about the only thing they are capable of doing. We have had six operators on my job during the past year. I wonder why? The general opinion is that they do not care for the clerical work, and still another reason is the town I suppose.

Its not much use for me to keep talking about being homesick lonesome and broke as that is getting to be the same old battle cry. That is generally understood.

From my window I now see a group of prisoners marching up the line guarded by two guards one in front and on in the rear. They get quite a lot of manual labor out of them all right.

Everything is going along fine with me and I am just about acquainted with the job, only I am rushed to death. Hogan has been out sick for about a week tomorrow. There was a lot of life in this ere town last pm. A couple of dances, wrestling match and three picture houses. A coon tried to murder a sailor too shooting up the main street but the police got them cornered. I am going to buy a

```
mileage book about the first thing I do as the
fares from Boston to Portsmouth are quite dear and
mileage saves you about forty cents a trip. Lots of
love and kisses till I see you dear.

                                    Herb.
```

The Southery, *a prison ship. After the Portsmouth Naval Prison opened, it became the station ship for the Navy Yard. Postcard view, c. 1908.*

The rumor mill must have been in full swing at the Yard that day. Herb's reportage of a "coon" trying to murder a sailor is colorful but somewhat at variance with the no less colorful account given by the *Portsmouth Herald* (an afternoon paper) on the same date, March 24, 1914:

> There was considerable excitement created on Daniel and Market streets shortly after ten o'clock when four revolver shots fired at a fleeing man, caused a great crowd to gather. The shots were fired by Officer Condon to stop Isaac Evans (colored), who was accused of an attempt to hold up a sailor named Ballard A. Ledgewood from the U.S.S. Southery. [The Southery, formerly a prison ship, was at this time, permanently assigned to the Navy Yard as a station/training ship.]
>
> The story of the affair as told by Ledgewood. was that he with another sailor, Lewis B. Kirby were in a lunch cart [Gilleys] with Isaac Evans, and Evans left before they did. Ledgewood and Kirby started down Market Street to go over to the navy yard when they met Evans, who they claim, held Ledgewood up and demanded his money. Ledgewood swung on

Evans and launched him down and then they started for the police station.

Near the station they met officer Condon and they stated that a colored man had held him up and with the officer they went back up Daniel street and Ledgewood pointed out Evans who was standing on the corner. When he saw the officer coming he started to run and to stop him officer Condon fired a shot in the air, thinking it would stop him, but it did not. Evans simply put on more speed and shot into Market street. There were more shots were fired, but they had no effect on the fleeing man, who ran through Ladd Street and was overtaken near the National Hotel and captured. He did not put up any fight when the officer caught him.

At the police station he denied having had anything to do with the matter, but was locked up. Edgewood and Kirby were held as witnesses but later released to Chief Master at Arms, who promised to have them present in court today.

Chapter 7

Portsmouth N.H.
Mar 26, 1914

Dearest Gracie Belle,

Your letter containing the newspaper clippings read. We are having great weather and the town is better with the exception of amusement. Mrs. B. is a fine cook and hard worker and has the loveliest little girl 3 years old that keeps her busy. They get up about 530 every AM & go to bed about 10 or there-abouts. The town is full of Naval officers in their uniforms which is quite conspicuous in all the theatres and public places. You speak about my new suit. I expect to get it next week as soon as pay day and if I stay I, in all probability will not be home till a month is up. Wait till you see this town spike. Just one view and you will be consigned to the hospital for one month. Its another story to leave home and have to be tied up in a burgh like this, believe me. You say the salt air will do you good. It will brace your appetite until you become acclimated, then you will know no difference. Mrs. B. said it took two years for her to make friends among the township. She says, "This crowd are the stiffest she ever saw." Fare to Boston 1.40 reg. tkt - 1.12 mileage. I'll get a book. Much obliged for news clippings. I can see Frank Jone's from here. [One of the five breweries in town. This letter was written from Waldron's lodging in Orchard St.]

Had a letter from Bill and he's going to New York to work as previously planned. He says not doing much at present in Bridgeport. You can see the Roosters and Chickens from my window. Tell Sis. it is a fine place for a chicken financier.

Speaking about the climate here, it is dry air and has no effect upon my head or nose. From home you cannot even smell salt and only occasionally noticeable on the river. Depends on wind. Sorry to hear Auntie has a cold. Love & kisses.

Bert.

Portsmouth N H,
Mar. 29, 1914

Dear Gracie,

Recd. cake and candy ok. Was glad to get both, for which I thank you. We have had all kinds of weather here and today is genuine March. We have worked one short at the office all week and I, being a new man not familiar with the work, made it pretty hard on the chief and Faust. Faust has been rather sore cause he loves work so much, and never hesitates to call the chf. lazy behind his back. I am not competant enough to index letters yet so cannot help out much, but occasionally copy letters and file them. In talking over saleries of telegraphers Mr. Young quieried me as to their average and if I thought the reason why oprs left the govt svc so frequently was on account of their betterment, and of course I did not deny that they could advantageously do such a thing, but kept right on with the bluff boosting average wages earned up as high as would look well. He did not say for whose information this inquiry. I dug out some interesting dope from the files the other day regarding the three names which were certified to fill my present position, giving the average percent etc, and also a lot of dope concerning my appointment with the wages paid all previous operators. On the list of three to be certified I was highest (not to brag, but this is the usual custom in selecting names) and the other two appearing hailed from Washington and Brooklyn respectively. I really think it worries them. There isnt much news except this is an awful town to live in. The correct pronunciation of Commandant according to the dictionary is Kom man dant.

Sunday I may go over to the yard a couple hours. Not much extra work to be had here. Lots of love & kisses

Herbert.

Portsmouth N. H.
Apr. 1st 1914

Dear Gracie,

Letter rec'd this am. Uncles letter recd, and candy for which I thank you, was acknowledged in my last to you. I am enjoying myself immensely going to the pictures every night in the week. Not much news to write so you probably wont hear quite so often as I am out growing my homesickness. I havent got thoroughly onto the job and probably will be a few weeks yet before familiar. I have no business telling you all of my troubles as you have enough of your own no doubt.

Although Mr. Hogan is irish I like him sociably the best of the three in the office. We go around the yard a great deal together, but nights there is very little company. I stroll into the Postal occasionally to see the operator & then go to some picture show. The opr from the Tacoma is going to give a ball overtown and says he wont forget me. I dont know what you do at this ball or how to act but Im going to take chanst. We had methodist minister to supper tonight and will have him right along during conference week. He is a fine talker and comes from the back woods of N.H. He left 2 ft of snow behind him. B. just passed off some April fool candy on me and it had some red hot pepper in it. Guess thats putting em over some. It keeps Mrs B. on the jump with me as a boarder, one minister and father & mother in law to feed and the kid to put in bed. She is nice though. She cooks up some fine desert and bakes all her bread and pardon me for bragging about an unknown person to you but her cooking is very, very, fine. When I landed here she didnt intend taking me as a boarder as it was new work for her and I dont know how she happened to do it. I guess Clarence must earn about 5 or six bones a day. He has been here 12 years. Best love & kisses

Herb.

A short article in the *Portsmouth Herald* for Monday, March 30, 1914, announces:

THE METHODIST EPISCOPAL CONFERENCE
HERE THIS WEEK.

> The annual Methodist conference which will be here this week, beginning on Tuesday will bring some three hundred clergymen to this city, in addition to the others who will be naturally attracted by the conference. To look after all of these delegates will be the duty of the local society that is to provide them a place to sleep with breakfast and supper, and it is no small job. All of the members of the society who have the room will entertain one or two of the delegates, but there are not enough to care for all and several people of the other denominations have agreed to take care of delegates so that they will be accommodated...

Chapter 8

Portsmouth N.H.
Apr. 2, 1914

Dear Gracie,

Your letter today not received. Have had a busy day today believe me. We undocked the Tacoma and Leonidas and now we have nothing left on the yard except the Mars [a coal carrier] and she is not being overhauled. Nothing new scheduled for dockage so this will be a dead place. The ships crew hate to put up in this hole and thats one reason they dont send more here. The captains prefer Boston or New York. One more draughtsman quit because his wife dont like the town. Hogan says this yard doesnt amount to much anyway, in the line of pay as compared to the others. He comes from Boston and thinks its about time he went back home. I believe you owe me two dollars, do you not. When may I expect to get it? I am not buying my $10 worth of mileage this time and may later on. Its pretty soft for you. How is the pillow getting along. Be sure and don't send it as I have no place to put it except on my bed.

They probably will be laying off some men before long. They can do that you know. They have been talking about closing up the yard and abandoning it altogether using New York in place and also talked of moving to Narra bay. So you can see theres nothing sure in this world.

Yes we had some more snow tonight. The ground is white and the fog horn, she blow. I guess this is some town. Regards to the folks. Lots of love and kisses.

Herb.

The *Leonidas,* which had been a collier, or coal-carrying ship, served the Atlantic fleet along the eastern seaboard and the West Indies off and on from 1898 until 1912. In May, 1912, she was placed out of service at the Portsmouth Navy Yard to be refitted as a survey ship. She was recommissioned on April 1, 1914, and immediately dispatched to survey the coast of Panama, probably in preparation of the upcoming opening of the Panama Canal.

The collier USS Mars *had been decommissioned and laid up at the Navy Yard, but on 8 May 1914 she was recommissioned and six days later dispatched to Vera Cruz. Postcard view, c. 1910.*

Chapter 9

Portsmouth N H.,
April 5th-13.

Dearest Gracie,

Its Sunday night again, and I spent the worst day of the week, which seems a week in itself, alone. Gee its a tough world living alone. And especially in a non-civilized country. The weather is not quite good enough to go to the beach yet so have to do the next best thing. Ever since I have been here there has been one clerk out at the office so I have to pitch in & do a little typewriting and letter indexing. I dont do any more than absolutely necessary for the more I do the more they expect me to do. This gang up here are the greatest bunch of stallers, even the Chief Clerk is called lazy by his understudies. When he comes in our office and leaves work the fellows all sit near the windows with their chairs tipped back, feet up on the radiators and answer his query with a yes, and then wait for someone else to do it. And they get away with it too. But the days pass so hard & long for the job is so soft it makes you keep track of the time. Hogan, a stenographer, who sits right opposite me says he often thinks of firing the job on that acct. & moving back to civilization. He has been here only little over a year and was transferred once to Guantanamo, but couldnt stand the climate. As soon as Faust gets back Young is going to stay out to fix up his chickens & he says he is sick with maleria.

Well Spike I know some one who is terribly lonesome and homesick for Aunties nice chocolate cake. You know I don't like to admit it though I can't come home before next pay day very well as I was so far behind. We get two holidays this month. Fastday the 17 and another both observed in Maine. I couldn't come home very well till my month is up as I would be unable to get a day off next to

Sunday so as to have two days & a quarter together. How is Grandma and the rest of the family? The wireless men are all fine fellows & every time I'm over there they invite me to stay all night & give *me* a feed. Two of them are going to buy a nice motorcycle. How does that sound to you. I guess they are always broke for I see em paying Hogan money quite often. The ball took place last Fri night but the originator and I got stung as some of the other fellows on the Tacoma took the whole affair in there own hands and did not invite Bruso so he did not go. He considered it a mean trick to do.

Eggs are selling for .20 a doz. Does it pay to keep hens? Ask Sis. Will close now with best love & kisses. I expect to receive a letter Monday am so they will pass each other enroute.

Herb.

Portsmouth N.H.
Apr. 7th, 1914

Dearest Gracie,

Yours recd. Monday am. Things going on about the same. We get a holiday the 16th but no pay with it. Pretty soft eh. Now I dont even know how to spend the day when a guy is broke. They have just revised the train schedule and we get three trains a day one leaves at 1:43 pm and the next express at 7:35 pm. Some service. Je spike you ought to be glad your free and can live in civilization but you can never appreciate those things till you endure married life. Lots of the prisoners are married and perhaps thats the reason they are in prison. There are but two movie houses here now and its hard work to fill em at that. In the Music Hall we have two vaudeville acts thrown in between which average fairly good. The conference is all over and the ministers all gone. Its been raining and snowing all day today. Mr. B. is only a clerk here & doesnt make but 3.50 per day. He has been in the service 12 years & might have been tfrd. to Washington with better prospects if he had tried real hard. Im working the stall game to the limit and maybe will be able to get off without doing much clerical work. They always try to make the new man the goat by passing all their work off on him while they sit with their feet on the desks but it dont work on me. I dislike to the extreme to do letter indexing as it is monotonous and rather complicated to learn. Young saw me practicing writing my resignation yesterday and ever since he has been extra nice about helping me to like the job. Now I can understand why none have been satisfied with their jobs. I thought I would resign yesterday but after thinking it over decided to wait a while as I needed the money bad. I think I will go down to Salvation Army meeting tonight. No where else to go.

I got Sis's letter this am and enjoyed reading it very much. She says she will be up to visit me this summer. Tell her we will be glad to have her but will have to get an extra bed unless I sleep down by the furnace. Remember we are ten cents distant from the nearest seashore and the only water to be seen in town is the Piscataugua River which I cross daily. York beach is .30 & Rye is 15 cts. Hampden is .10. And after you get there you cant go bathing. Well Gracie, I expect to be home Sunday the 19th. Love & kisses.

Herb.

Portsmouth N.H.,
April 8th-14

Dear Gracie,

Your letter with Aunts card recd for which I thank you both very much. 5 Pm package not yet recd. but I know it will by tomorrow Am. It certainly will be good to get back to civilization once more. But its pretty soft for me just now. Feet stuck up on the radiator, cigars, and nothing to do till am. I can sit there and worry my head off. But I have no right to worry. Im not married yet and when I get married my wife can do the worrying. There will be a lot of clerical work to do when I can make up my mind to do it. The way this office is run, you are supposed to do everybodys work with your own & then they are all happy. In other words there is enough work to keep one busy all day. Well why talk about work. Its supposed to be perfectly proper & customery to work for a living. There are four office boys from different office buildings viz. Constn. & Repair; Supplies & Accounts; Steam Engineering; and the accounting Dept; which hang out a great deal of their time over in the back room from my office and shoot dice for pennies. They are a circus to watch. Their names are Crowley, Williams, Long and one more who has a name. Some amusement.

I rec'd lots of cards today with lots of reading. Haven't heard from Alice yet. Went to the movies last night and saw a fine show. Two hours and half long. Mrs B. is entertaining more company her husbands brother & wife, father & mother-in-law.

Notice the class to this writing paper. The reason I use it is because it costs me nothing. No, the office furnish it Or else Dywer left it. More things to worry about. I dont think I have grown fat any since my vacation began. I never have been in Frank Jones's yet thats the reason. I asked Hogan why he did not get married and he said it was no place to bring up children in this burg as they never would have the educational advantages to prepare themselves for their future livelihood which is almost correct.

Tell Uncle I enjoy reading his book very much & give my love to all & tell Auntie that I am in hopes of eating some more of her cake & remember that Im not always the good little boy.

Yours

Herb.

Portsmouth NH.,
Apr. 11, 1914

Dearest Gracie,

This time I will answer your latest letter am going to mail it Sat night so you will receive it Monday morning. You pass a half unkind remark when you say you dont know as I will be glad to see you. You know better than that and when you say Hogan and I are two rascals, remember and dont blame me for what other people say. Im not responsible. Lets see tomorrow is Easter Sunday, and if I cant find anything good to do I think I will go to church.

Speaking about chickens this is some chicken town. Enuf said. Lots of motor cycles on the streets.

I went over to the dance hall Wednesday night and saw some very nice pictures (Gen. Film Co.) but there is not much class to the audience. Sunday again. Je I dont know what to do on Sun. But by next Sat. night late I expect to be back home if I can make connections at Boston but if I miss I spose I'll tie up there rather than arrive home at 320 Am in the morning. Young is planning to be out some time this week and that may knock my plans but I expect to do it if possible. The wind blew about sixty miles an hour today shaking the windows hard. Thanks much for the Easter Card. You must come up and give the town the once over when the weather gets good. Mrs. B. is sick and so I dont know as she will be able to keep me much longer. So I darst not ask her to fix you up yet. Would you come if she takes care of you. Theres no reason why you couldnt some Sat. pm & stay over till Mon. Am. She is going to Boston Wed. to stay till Monday. Love & kisses.

Herb.

Portsmouth, N.H.
April 15th, 1914.

Dear Gracie,
SUBJECT: Coming Home.

The chief is away on a leave of absence. and the outlook is quite gloomy, although I am not an absolute necessity to the office they dont care much whether I am here or not, but it looks better to have me around when short. As tomorrow is a holiday I thought I would beat it then for home but they failed to pay off today so will be obliged to wait till Friday for my pay and that will mean Saturday night for my home coming, if nothing further turns up. Again should I prove to be broke on pay day, I could not make the contemplated trip, and so would add another bunch of disappointment to all concerned. You must remember that I have only been here approximately one month and you could not expect your angel to become a million-air in that short a period, as it costs money to live, pardon me, exist. I cannot live on my present salary. Dont have a blooming thing for excitement around this burgh, so am seriously thinking of going up to Dover tomorrow to witness the boxing match in the evening, and the ball game in the afternoon. This is a city about the same size as Portsmouth. The whole state of N.H. hasnt as many people population as Meriden Conn, has, so you can easily see that this is some proposition to buck up against. Mrs. B. couldnt go to Boston this week as her husband would not allow it, and she was terribly disappointed, and they had a regular family scrap, and I thought what a misfortune to be married like that.

SUBJECT. FERSON AND HIS MOTOR "CYCLE". Ferson had a big spill from his machine the first day out and nearly broke his neck. On the second day while riding 75 miles an hour struck a mound in the road and landed way over in the field breaking his machine to bits, so he left it along the wayside in some garage, and after forgetting which one he left it in returned home via the B. & M. train using the money he had previously borrowed for gasoline to pay his fare with. The next day Baker had to go up to York Beach and hunt around for the garage and finally brought the machine home. Now they have exchanged it for a new one.

Love and regards to all.

Herbert.

April 26, 1914.

Dear Gracie,

Its Sunday again in this burgh and it has been raining hard all day and I have stayed in but you can bet I wish I was coming over to your house tonight. Things are moving over at the office and all the marines are leaving and that includes our orderly in our office, which will make more work for us to do. According to the papers war is over, but dont you believe it. Should this be the case just think what a disappointment it would be to our Army and Navy. I suppose we will have to work short all summer now just to give the boys a good time maneuvering around. Je Spike you should be glad you have your Auntie to look after you these dark dismal days and then she can't be homesick. Ha! Ha! I guess you will see me home before long. Je, its tuff to live alone. You cant live in this burgh, you merely exist. Ive been away from you a week tomorrow (Monday). Mr. B. has gone down to the armory to receive volunteers. He's waiting for me.

Im in an awful predicament besides being alone, I is smokeless and candyless and we dont have any supper tonight and I have no girl to love - I have sent you a little package by parcel post to sort of settle up my bets with you and make you happy and lonesome, although you will be lonesome tonight with or without it.

Faust is a rather mean disagreeable chap to work with, no one cares for him. I have to go in and confer personally with Capt. Field, Commandant acting, occasionally now that Skully has went, in reference to all telegrams. In all probability two wireless men will be obliged to go. 140 prisoners fm Boston yard were transferred here. How do you like my new suit? You should see the combo Young wears, a so called installment suit. He must have had a hard winter. I might make this letter longer but whats the use it only makes me more homesick. Lots of love and kisses.

Herbert.

P.S. If you get too lovesick send for me and perhaps I'll come -

This is the first time Herb mentions the war and associates it with the Yard's problems,though he has, throughout his letters, alluded to the lack of work at the Yard and the constant threat of layoffs or a total closure of the facility.

One of the prime reasons for a dearth of work was the fact that the entire Atlantic Fleet was engaged in an invasion of the cities of Tampico and Vera Cruz, Mexico. On February 13, 1913, Mexican Army Chief Victoriana Huerta had Mexico's president assassinated and took over the country. President Wilson refused to recognized the new government and actively supported Huerta's rivals. Wilson, on the pretext of assisting three U.S. sailors who had been temporarily detained in Tampico by the Mexican government, ordered the entire North Atlantic Fleet south.

Thursday, April 16, 1914, the *Portsmouth Herald* carried the following above-the-fold headline and story:

> NORTH ATLANTIC SQUADRON SENT TO TAMPICO
>
> (Special to the Herald) Washington, D.C., April 14.--Secretary Daniels issued orders at 2.30 today directing the entire North Atlantic squadron proceed to Tampico, Mexico. The Tacoma at Boston, and all other ships there, ordered to rush. All the marines at New Orleans directed to embark immediately for the scene. Ships under way and bound north have been intercepted by wireless and ordered to return. The administration has determined to enforce the salute to the American flag and possibly intervention. [One of Huerta's sins was to refuse to salute the American flag on arrival of visiting U.S. Naval ships.]
>
> The torpedo destroyers, seven in number, at Pensacola, and others at Charleston, have also been directed to proceed to Tampico and report to Admiral Fletcher. The department has also directed all the ships at New York to sail immediately. It is expected to have the entire North Atlantic with all auxiliary ships there at the earliest possible moment. The orders have created a sensation at the Capitol and old war-time scenes have been enacted at the department. The president's special representative, Mr. John Linds, report it to have been the cause of the change in policy.

The entire squadron was tied up in Mexican waters for the next two and a half years. Scheduled Yard overhauls were postponed or canceled altogether and, presumably, any emergency repairs were done at southern yards much closer to the front line. Though there were many serious incidents with loss of life on both sides, all-out war was narrowly averted by diplomatic means and the squadron was withdrawn in January, 1917.

Chapter 10

April 30th, 1914.

Dear Gracie,

Today is Thursday and I am feeling somewhat better as it is rapidly approaching pay day, and although my roll will be somewhat short on account of holidays, etc. I expect to survive the shock. I have been feeling rotten ever since coming back this last time, and dont know whether to lay it to the water or climate. We are having terrible weather here, rainy and cold and there is nothing to do in the office to keep me busy except to do everybodys work and this makes me sore at times as it is an imposition.

For a wonder everyone has been on the job this whole week, excepting the orderly who went to war. I will have to take my share of that work, which by right belongs to a kid. Portsmouth society is very cold and distant, punk as it were, and it is very difficult to find suitable fellows to chum around with. Mr. B. spends all his time at the Armory getting recruits, every night and Sunday and leaves his poor wife all alone at home. She is hollaring for a night off but he says nothing doing this week.

This noon Hogan and I went for a walk over to Kittery. None of the other clerks seem to venture out of the yard at noon and I have found none that I personally like to chum around with. I go in to visit the fellows in the Western Union, and also the Postal every other day or so. One of the W.U. operators here tried for this job some time ago but only stood 662/3 which is very good for a fellow of his experience and character. Everything is real quiet here and not much news from the war path. The

Commandant is the only live one around here, and he keeps me jumping when there is not much else to do. He seems such a nice sort of a chap, but is not suppose to come in contact personally with his clerks. I have been contemplating taking out sort of Insurance more than ever recently as I am expecting to die most any time if I am obliged to stay here much longer, and I will make you my beneficiary. You undoubtedly will get enough money to bury me with. I havent decided yet what company to tackle, but have an idea it will be the Postal Life of New York as they offer the most attractive policy for the least premium money. If I dont feel any better in the future than at present it is about time I got busy with some kind of a policy, is it not? Yes. It is not. To be continued next month. Guess I will close now, with regards to all the family. Love and Kisses.

Herb.

May 3rd-14

Dear Graciebelle,

I am beginning to realize what a great number of letters I have written you the past week & can see that I am spoiling you for so do, but however I have but myself to blame. By this time I presume you have become accustomed to being alone, that is with the folks alone, which is not as bad being here alone in the world with no mother to guide him. Next comes the weather It is a beautiful day but cool. Things are quiet at the yard. No ships in but the Mars & they have had orders to work on her day & night. Rear Ad. Rogers Cmdt, returned yesterday and was retired. They fired 13 salutes for him which shivered our timbers for fair. The suffrigettes had a parade in Boston yesterday. Mrs B. had a fine dinner today and we had 2 kinds of desert; mince pie, and Maggots [a colloquialism of the day for rice pudding] with banana & whipped cream. Have I your permission to go to war? If Mr. B goes & I stay here, and his wife goes to her home I will have to look for another boarding place. That will be tuff luck, eh.

My new shoes look real good but the soles are wearing down at the toe and I am getting a flat foot, right one, on account of walking so much. It makes my limp quite bad at time. I never ride now. You cant get Young to walk home on a bet. You can bet I miss the folks & Hfd very much, and you the most of all. My toncilitis is much better now. Mrs B has a bad boil (I dont mean water) on her neck & the kid was sick with brain fever or something & the old man had a sore throat. Some climate. Lots of love & kisses.

Herb.

Portsmouth, N.H.
May 4th, 1914.

Dear Gracie,

Your Sunday letter not yet received but I expect to get it this afternoon on my way home. I was so homesick Sunday that I did not know what to do and so after dinner I went to bed till 5:30Pm, and then went down street to mail your letter and returned home for the night. I feel pretty punk lately and guess I will have to cut out dissipating so much as it does not seem to agree with my makeup. I think it is the drinking water or else my diet that makes me so groggy , for I never do any work for excitement. Maybe that's what the matter.

We are going to have minstrel and dance this evening over street, and I bought a ticket at $.35. Hogan says the best way for a single man to spend Sunday, and he talks from experience, is to sleep till noon, without any breakfast, and then get up and get replenished, take a walk, and then turn in for the night. The yard will be closed down by the summer months if they dont call this war stuff off before then. They will lay off a lot of men for a long period as there is no work for them to do. I dont know what will become of the office force, but am inclined to think that they will be effected also. I am writing this to spoil my little darling so she will expect to hear once daily. But why should you worry. I often wish you were here with me and then I am glad you are not at present, for it would be such a bore for you to hang around this place with nothing to do to kill time. It would make you so unsatisfied with life that you would go back home quicker than you came. I am not yet satisfied with the place.

The sociability here is a freeze and the town is a frost. Mrs. B. will vouch for that, after serving twelve consecutive years but she did it for her husbands sake and now she has got to put up with it whether she likes it or not. She is planning to go down to Putnam for a rest before long, but I doubt if she ever gets there. I suppose you went to church Sunday as usual. Well you know my failings. Mr. B. is going to get some nice young lady to call for me on Sunday and take me to church. Wont that be nice?

There is no special news, but please give my regards to all the family, and love and kisses to yourself.

Herbert

Waldron's minstrel show was put on by the Rebekahs, a women's fraternal group closely allied with the International Order of Odd Fellows, at the Freeman's Hall in Portsmouth. The *Portsmouth Herald* for Tuesday, May 5, 1914, gives a glowing review for this amateur production and reports "there is still another (performance) to be given this evening."

Minstrel shows were a quintessential American entertainment. They were inspired by the southern plantation slaves' songs and dances. The creation of the first minstrel company is attributed to Thomas D. Rice of Virginia, who started touring with his company in 1830.

The minstrel show had a highly structured format that remained constant throughout its history. The orchestra was at the back of the stage with the players arrayed in a semicircle in front of the band; a group of six "end men" were placed in the very front with the master of ceremonies, or "interlocutor," in the center. The entire troupe except the interlocutor were in "black face" (black grease paint); the interlocutor was in "white face." Jokes and snappy repartee in black and other ethnic dialects were exchanged between the end men and the interlocutor, interspersed with musical performances by individuals and the entire troupe. Throughout its tenure, the minstrel show

This 1860s book contains enough skits to stage several minstrel shows. Included are "Professor Unworth's Celebrated Lecture on the Atlantic Cable" and "The Dutchman's Lecture on War."

maintained its format. The material was constantly updated to reflect local and national current events while retaining some of the old standbys.

In the 1860s one could buy *Brudder Bones Book of Stump Speeches, and Burlesque Orations...*, which contained enough material for several minstrel shows. Among its offerings were "Professor Unworth's Celebrated Lecture on the Atlantic Cable" and "The Dutchman's Lecture on the War," which refers to the U.S. Civil War.

Initially these shows were presented by professional touring companies of all male actors. By the turn of the 20th century the minstrel show had been adopted by amateur groups to provide community entertainment and as a fund raising venue. The prohibition against female actors seems to have also gone by the boards. The *Portsmouth Herald* review of the minstrel show Waldron attended tells us that not only were several of the leading players women but the interlocutor was also female; all were judged to be extremely competent at their craft by the reviewer.

The main knock on minstrel shows was that they were racist, and that is undeniable. However, during their time, minstrel shows were attended by equally enthusiastic black and white audiences. They were the foundation for burlesque, vaudeville, and finally the American musical play and movie. In the beginning the minstrel show was the exclusive purview of white actors, but later on Negroes were admitted to these companies, though they also had to perform in black grease paint so there would be no variation in shades of skin. It's likely that minstrel shows were the first place that black entertainers were given experience and exposure on stage.

Minstrel shows were a popular American entertainment for well over one hundred years, starting in 1830 and lasting until the 1950s. Their decline was already under way when Herb went to Freeman's Hall. The movies and then radio and television took their toll on the minstrel show, as they did on most other forms of live entertainment. This author can remember attending a full black face minstrel show when he was a schoolboy in the early 1950s. The Civil Rights movement, beginning with the end of that decade, sounded the death knell for the minstrel show. In the late 1960s, a local community group in Kittery, ME, presented a show called "The Barnyard Frolics" at the Robert Traip Academy. The format was straight minstrel show, but the actors had been stripped of their grease paint and all the jokes had been stripped of their ethnicity and given a more generic, more politically correct "country bumpkin" façade.

Chapter 11

May 5th, 1914.

Dear Gracie,

Please address all future mail to 48 Orchard St., instead of General Deliver, and anything important to Navy Yard. This is necessary now according to the Post Office laws. Love and Kisses,

Herbert.

May 8, 1914

Dear Gracie,

I am still here on the job and we are having so much bad weather that it makes a fellow homesick. I am feeling fairly well but everybody seems to complain of colds and sore throat here account of dampness. By the way my name is in the directory [Portsmouth city directory, ed.], so I guess I must dwell in this city, or I should say country for we live, excuse me again, I should say exist on the border of the country. Its just a good brisk 20 min. walk from the house to the dock & five mins. from the other dock to the office and 5 mins across the river so there goes 40 mins. easy from here to work. And by the way I have worn out those nice tan shoes and had them resoled at $1.00 a throw. Nobody rides here. But figure it out. It certainly cant be cheaper to walk at the present cost of shoes. I think Im getting a stye in my eye. Some troubles Herb. does have. Again speaking of the country, Spike, we have music every night, such as the sweet voices of the bull frogs and the roosters crowing and besides the air is so sweet here. Its a lovely place to die in. My appetite has been normal since my return the last time so you can see my previous statements herewith proven. The Smith girl is laid up with a terrible cold-cough. Commandant Acting Field is very nice now and he dont bother me but once a day. No commandant has yet been appointed to the yard. The Montana is coming next week and the Wheeling follows. It looks as if they were trying to weather the work thru if possible. The Mars sails the 14th. Love & kisses & hugs and everything.

Herbert

Either the rumor mill was still in full operation or some more availabilities were canceled; records indicate that the *Wheeling*, a gun boat, did not visit Portsmouth during this period. If it did come into port it must have been for a brief refueling/ replenishment stop that didn't warrant a mention in its war diary. The *Montana*, an armored cruiser, did come to the yard and was dry-docked for repairs.

USS Montana. *Postcard view, c. 1901.*

Chapter 12

May 10th 1914

Dearest Grace,

Everything here is going along as usual, this town only allowed three things , morning noon and night. There is always a scandal about town and always somebodys next door neighbor telling em what they will & wont do. The town is still unconscious although its past reputation has been the worst, it has eleven churchs to fight satan with. Down street Sat. night there were about a thousand people on the gay white way. "The perils of Pauline," one month since released, played here Sat. [Herb was witness to the birth of a cliché!] I have been here nearly two months and havent seen a beach yet and they are tucked away in such places that there isnt much danger of seeing 'em either. Its beautiful weather today warm etc. but the nights are so cool. Just had my shoes tapped and they look fine now. This is sure country life wit salt air on the side. Nothing to do till tomorrow. The operator at the RR depot lives in the other side of the house and he took the Civil Service exam for RR mail clerk. Pierce hall (movies) suffers greatly since the Marines and ships departed from the yard. They have a new floor mgr; the other resigned to do nothing for a living, and this is a bad place to hunt for a livelyhood most persons living on their interest of what they owe. Well Spike its only six weeks now till I see you again. Not a chance to get any extra work here as there are 5 extra men available at all times & they dont pay enough to make it worth while. Very much love to all. Kisses to you.

Herbert.

Tuesday, May 12, 14.

Dear Gracie Darling,

Today being Tuesday and a very dull day, as the whole force seem to be on duty, it affords ample time to write you a few lines, which will not be necessary to do on Wednesday, the Yard is real dead now and excepting when some one stays out there is no work to be done here. They are going to fire three draftsmen the First of July rated at $5.50 per day each, as there is no work for em to do so guess they will have to beat it out of town. This always happens in some branch of every Navy Yard at the end of their fiscal year, as they usually run short on the appropriations. This year it happened to be in the Public Works Dept. When it occurs in my dept. then I will get the hook. Its a funny stunt they have of pulling off in this branch of Govt. work and that is one reason that the Navy Dept. is not a desirable place to work in.

Pa wrote me that Mgr. Ryder was to retire the first and that bum Russell would fill his place. I guess there will be some class to the W. U. when this takes place. I am glad I dont work for him. Pa doesnt like it very well. The Mgr. of the W.U. here is very rarely in the office, and he has a graft almost equal to mine. His wife is an ex and she comes in to look after his business occasionally, but her real profession now is proprietor of the swellest millinery business in town. He has an auto and is rather an old man and he is very deaf. He has a daughter about 21 years old, who went to New York to study physical culture, but she did not like it so she sent around selling adding machines. That sounds like a funny stunt for a girl to do in New York especially coming from this place.

There is a nom new scandall over town at present. No news of somebodys wife leaving town misteriously. We are having rotten weather just now if any body asks you. Fersons wife is due in this burgh the first of the month. He is having a hard time of it financially I guess. Mr B. is still on the job here, and does not seem to know anything definite about getting started. So I guess he will be stalled off with his wife. That will be a disappointment for him. Your Sunday letter just received Monday Pm.

I still eat three meals a day, but it is awfully lonesome here for me. I manage to kill most of the evenings in some moving picture theatre, Business is just beginning to pick up now when I get settled down for a rest. But why should I worry? I have something to look forward to you know. Yes Young informs us he will be out tomorrow. More work for the weary. Je how I hate to work. You know my failing. I get awakened every morn. at 6 now, as Mrs. B. and the kid are generally howling about that time and there is no sleep for the lazy, when once the kid gets on edge. Well love you for this time, and send lots of kisses to my darling.

Herbert.

Waldron's first paragraph mentions E.M. Fisher, manager of the Western Union office at 22 Daniel Street and describes his wife as an "ex", meaning she was an ex- telegrapher and that she sometimes came into the telegraph office to handle some of her husband's supervisory duties. Although she may have been exceeding her authority in this instance, women were commonly employed in the telegraph industry, and a few made it high up in the ranks of middle management before the beginning of the twentieth century.

The first woman to be appointed to a position in the fledgling telegraph industry was Miss Sarah G. Bagley, who was appointed superintendent of the Lowell, MA, office of the New York and Boston Magnetic Telegraph Company in early 1846, less than two years after Morse's first public demonstration of this new invention. Bagley was a Lowell "mill girl" and an early woman's rights activist. The telegraph appointment was more notable to her contemporaries for the crossing of class rather than gender barriers.

The next woman operator was Phoebe Wood, sister of Ezra Cornell. Cornell made his fortune in the telegraph industry, having started working for Morse on the first line from Washington to Baltimore. Wood became an operator at Albion, Michigan, in 1849, having been appointed by Cornell's partner, John Speed. Speed also appointed a Mrs. Sheldon to take the Jackson, MI, office. Cornell was in favor of these appointments, saying that at low-salaried, small stations would likely get better service from a woman. The women could employ their idle time sewing or knitting.

J.D. Reid in *The Telegraph in America* reports the next two appointments. In 1851, Miss Emma A. Hunter was made operator of the Westchester, PA, office (near Philadelphia), at a salary of twelve dollars a month, with a year end bonus of one hundred fifty dollars.

In March, 1852, Mr. Charles F. Wood appointed Miss Ellen A. Laughton, age 14, to the management of the Dover, NH, office. Wood comments on Laughton; "When Miss Laughton, in 1856 was in charge of our office, at Portsmouth, N.H., I regarded her as one of the best operators I had ever seen."

So how does a young lady of fourteen become "manager" of a telegraph office? In Ellen's case, the short answer is nepotism. On the 24th of June, 1836, Eben (Ebenezer) O. Laughton and Serene M. Potter were married in a civil ceremony in Litchfield, ME. On July 1st ,1837, Serena gave birth to Ellen A. Laughton in Litchfield, the oldest of three daughters. By 1846, the family had relocated to Dover, NH, and Eben was a partner in the firm of Laughton & Whidden, Auctioneers, Franklin Square. Laughton had become the sole proprietor when this column appeared in the *Dover Gazette* for March 20, 1852:

> We are informed that our old friend Mr. E. O. Laughton has entered into an arrangement for the purpose of keeping communication by Telegraph open to our citizens, and has so far accomplished his designs, that he has already opened an office at his Store on Franklin Square, where he will be ready at all time to attend the forwarding of any communication, entrusted to him for that purpose. This enterprise in Mr. Laughton, we predict will meet with favor and support from our townspeople, and as Mr. Laughton has been at considerable expense, to connect the wires, and put everything in working order, we hope that he will be liberally renumerated. The citizens of this and the neighboring towns, more especially the business portion of them have long felt the need of the telegraphic wries (sic), to convey intelligence of various kinds, but no one had the courage or disposition to effect the desired object until Mr. L., volunteered to perform the work necessary for its successful operation on his own responsibility. All messages entrusted to Mr. L., the public may rest assured will be kept strictly confidential, and no effort will

> be spared on his part, for the faithful discharge of all duties connected with this branch of his business. Shall he be encouraged by the patronage of our citizens and those of the anjoining towns!

Ellen was taught to operate the instruments by a Mr. Frank Nelson, who was probably an operator employed by the Morse Line to set up new offices and train operators.

By 1856, at the age of eighteen, Ellen had taken charge of the office at Portsmouth; the family was residing at 26 Deer St. Meanwhile, Eben had become a major contractor of New England telegraph lines. As the *Portsmouth's Morning Chronicle* for Jan. 9, 1856, notes:

> Cape Cod Telegraph.
>
> The Cape Cod Telegraph Company which was chartered for the purpose of constructing and maintaining a line of telegraph from Boston to Provincetown, Holmes' Hole, Tarpaulin Cove, &c., is now completed, with the exception of laying a cable some five miles across Vinyard Sound. The first pole was raised in Boston, the 10th day of July last, by E.O. Laughton, of this city, the contractor and builder of said line, and it now extends over a territory of 225 miles, built with great care and best materials...

The article continues with several paragraphs of details on the line and which amounts to a prospectus for stock in the new company.

Ellen apparently accompanied her father for some part of this venture (perhaps to train new operators), as a column for Aug. 12, 1857, in the *Morning Chronicle* tells us:

> We learn that Mr. E.O. Laughton, the well known telegraph operator, has sold his interest in the Commercial Telegraph Company, at New Bedford, Mass.; and is again to take charge of the office in Portsmouth. Miss Ellen A. Laughton, his daughter, a skilful operator [or operatress], will resume her old position here, on the first of September.

It's unlikely the senior Laughton was an active operator and though he was probably a stock holder in one or more telegraph companies (contractors in the early days often took some of their pay in stock), he probably was not the "manager" of the Portsmouth office as this article indicates. In the several city directories where Ellen is listed as an "operator" of the American Telegraph Co. office at 13 Pleasant St., there was never any other operator or a manager mentioned. Some of the competing offices did list multiple operators and designated a manager. Also J.D. Reid's history identifies her as the manager of the Portsmouth office.

In April, 1858, Ellen joined the First Methodist Episcopal Church in Portsmouth. In 1859, a young mariner and Portsmouth native by the name of George W. Thompson also joined the same church by profession of faith. Eben Laughton died December 24, 1859, while in Boston on business. His body was returned to Portsmouth and he was buried Harmony Grove Cemetery, off South Street; he was 45 years old.

The Telegraph Office at 13 Pleasant Street where Ellen Laughton worked for nearly ten years. From a Davis Bros. stereoview, c. 1865.

On December 24, 1865, Ellen A. Laughton and George W. Thompson were joined in marriage by the Rev. James Pike of the First Methodist Episcopal Church (today's First United Methodist Church).. The couple resided with Ellen's mother at a boarding house at 56 Daniel St. Ellen's telegraphic career appears to have ended after over thirteen years of service in Dover and Portsmouth, NH, but her legacy as a woman telegrapher was carried on.

Ellen's immediate successor was a Miss A. J. Cole, who held the post until some time after 1872. Cole was in turn succeeded by Miss Maggie J. Nutter, who held the job from 1872-73 thru 1879. Sometime before 1881, a male operator finally took over the office.

Ellen's first child, Willis M. Thompson, was born in Portsmouth on March 9, 1868. By 1871, the family, including Ellen's mother Serena, had moved to New Bedford, MA, where Ellen gave birth to her second child, Henry L. Thompson. Ellen's husband George served in the Light House service, rising to the rank of Master of a light vessel by the late 1890s.

The 1880 census lists the family as follows: George W. Thompson, 43, head of household; Ellen L. Thompson, 42, wife; Willis M. Thompson, 13, son; Henry L. Thompson, 9, son; Serena M. Laughton, 62, mother; and Mary A. Mareema, 33, maid. The two boys followed in their father's footsteps, both becoming sailors. George died in 1903. The last listing to be found for Ellen shows her to be a widow in 1909. She was 71 years old.

Eben O. Laughton's grave. South Street Cemetery, Portsmouth.

Chapter 13

Portsmouth, N.H.,
May 15th, 1914.

Dearest Gracie-Belle,

Your letter received yesterday pm, and was glad to hear from you. Also the newspaper items with it. I thought I would send you a copy of the town newspaper as it would be of interest to you, although there was nothing about yours truly in it. Today is a fine day and there is not much work to do. Young and Faust were both out all day yesterday so that gave me something to worry about. There are no ships at the yard but the Montana is awaiting to be docked in the lower harbor and probably be docked in the morning. There have been several men furloughed in the machinery division for fifteen days without pay so you can see the future outlook is very poor.

Mrs. B. is going to Boston for sure Saturday and probably will be gone all the week, and I think the best way is to address all my mail to the yard for the coming week as no one will be there to take it. Perhaps you dont know what those initials stand for after H. D. Waldron's name, that G.P.A.M.C.R.R. Well if anyone asks you it stands for "general passenger agent Maine Central [Railroad]. I guess there is some class to me all right. I haven't got my life insurance and also the cheapest premiums on a $1000 policy. It costs $2 initiation fee, and only an average of $1.00 a month, payable only when some one dies. Or possibly some months it would amount to a little over a dollar. The Postal Life of New York does not make such a good offer as it costs several dollars more a year for the same amount of insurance. For a straight life the W.U. is the best.

I am not sure that I could pass the physical examination, as they are fussy who they take. One dollar a month means an average of twenty five cents a week.

Well Gracie I expect to be home about in 10 weeks from today. How does that suit you? I am buying a mileage book this week, as it is a necessity, saves one half cent on a mile. Now that the N.H. road has gone up on the fare I suppose it will cost me about fifteen dollars to make the round trip. I dont think you will get home very often after you once leave home. Ha.Ha. Thats a funny one is it not. You can see what you are coming to. I am the boss. Hey? Let me be the Boss for a while. I accept the nomination. Regards to all the folks and Auntie, and tell her I miss her pies and cakes and feed. Lots of love and kisses.

Herb.

The railroad and telegraph companies formed a symbiotic relationship soon after the first telegraph lines were strung. In addition to public roads the telegraph sought agreements with the railroads to string their wires alongside the tracks. Initially this was for the convenience of the telegraph companies and they paid the railroads for the privilege, but it wasn't long before the railroads realized that they could utilize the telegraph to schedule their trains and maintain safety on single track main lines. Agreements were soon reached that gave the railroads absolute priority for rail safety messages after which the lines could be used for press, commercial, and private messages on a first come, first served basis. This brought telegraph service to small communities that otherwise would not have rated it and the telegraph and railroads shared the expense of operators and equipment maintenance.

At smaller stations the station master, male or female, often served as ticket taker, baggage handler, freight agent, switchman, and telegrapher all in one package. A General Passenger Agent Maine Central Railroad (G.P.A.M.C.R.R.) would have to be a versatile person. Waldron ultimately did not assume that title but his letters do not reveal why. It's only speculation, but there are a couple of possible scenarios.

The Yard's lack of work was not due entirely to the Mexican war. Articles in the Portsmouth papers during this period speak of a business depression in progress. One article reports:

> New York, March 29. - The Pennsylvania railroad, in addition to 25,000 men laid off on its eastern lines, it was learned Saturday has cut down on its lines west of Pittsburgh by about 12,000 men, making a total of about 38,000 men for the Pennsylvania system as a whole.
>
> Yesterday the New York Central announced that since December it had reduced its force by about 25,000. These two systems have thus laid off about 63,000 men since they began their present program of retrenchment last December. The heaviest cut in forces was made last week...

The article later indicates that the Boston and Maine was bucking the trend and actually hiring more people, but the economy may have eventually forced them to reverse themselves and they may have had to rescind some offers of employment. It may also be that Herb had simply realized that he could no longer stand being away from his friends and family and most especially Grace.

Chapter 14

May 17th --

Dear Gracie,

Mr. B. & I are keeping house all alone and we are getting, I mean he is getting the meals for us every day. He is some cook and we are enjoying ourselves as much as possible. I am getting discusted with this place and when I come home I may never come back. My new suit as I told you before is a blue serge. It is thin, quiet in color and looks nearly black.

This town is full of scandles. Thats the chief topic on the streets. Heres the last one I heard. See how you like. Its not naughty. The piano player at Pierce is engaged to the singer, a young girl, but not married they both are staying at the National Hotel. They both earn $3 a day. News. Something to talk about. I hear 'em all spike . I heard that one at Postal. But there is still another one still better than that but I guess I wont tell you. Her name is Hope Walden a young lady of money and was married to a post office clerk, but not now. The Montana is a fine ship and here crew were giving the town the once over Sat. night, -. I will be home in just seven weeks from today. To stay - maybe - not - but perhaps yes - Shall I - yes - all right - I will. Disregard previous letters address all mail except Sundays to the house. Its pretty hard to tell you how much I love you right here as it makes me home-sick, but you can be sure that if I was there every bone in your body would be lame or broken - Im a sick man anyway. I guess I dont get enough sleep. The Smith girl is better now. She has a fellow on board one of the ships. Shall I come home anyway for the fourth of July. Lots of love & kisses.

Herb.

Portsmouth N.H.,
May 19th-14

Dear Gracie,

I received your letter addressed to "G.P.A. M.C.R.R.," yesterday O.K. Since Mrs B. has been away we are keeping bachelors apartments eating our two meals a day at home and doing the dishes there after. He is the cook and gets up early every Am & has the grub ready when I alight in the morn. Last night after supper we went to the ball game and after that we went to Pierce hall till bed time and then after reaching home we dove into some rhubarb pie, that our neighbor sent in, and had a fine time till 11 Pm when I thought I would turn in for the night. This morning we had a present, or donation of a dozen hot biscuits, right from the oven for breakfast. We have the promise of a short cake from another neighbor some time this week. Mr. B. says if this good stuff continues he dont care whether his wife comes home or not. There were 3000 turned to see the first ball game of the season yesterday and the players were accompanied by the yard band. The Montana is in dry dock now and she is a pretty ship. All the jackies had their photo taken in a panarama picture yesterday.

In the Public Works dept there are six draughtsmen employed and it has been planned to lay four the first of July at the rate of $4.50 a piece per day. Several machinists were furlough Saturday.

Well I expect to be with you in exactly 9 weeks from today.

Lots of love & kisses.

Herb.

Portsmouth, N.H.,
May 21, 14.

Dear Gracie,

I received your letter yesterday O.K. Was tickled to hear from you so readily. Things are pretty slow here as usual and we have run behind $70 on our alloted funds. Pay Misc. which means that some body in this office will have to take a months vacation without any compensation. Probably it will be me. Wont that just be grand. I think so. Well there is no use worrying anyway, as I will be home then. The Montanas crew have gone ashore on leave for forty eight hours, and they all carried souvenirs that they have brought home with them from the tropics, consisting of monkeys, parrots and birds. Should I get stuck for this months vacation it is doubtful whether I would receive any salary for the latter half of this month. It is some way of doing business all right.

Was just talking with an operator who has worked in Hartford at the Courant and he has been up against the rocks, and now is out looking for a job. Mrs. B. still away and we are eating out at the house, and doing the dishes ourselves. Except last night we went to the church for supper. Everybody goes to the ball games. The weather is bad as usual and it gives me a grouch. Je, aint you glad you are not around to hear it.

If I dont get that vacation pretty soon I suppose I wont see you for twelve weeks. But I am liable to be home tomorrow. You cant tell, what will happen. The fellows in the drafting room are going to get a whole years lay off. Four of them. Dont tell anybody outside the family about the hard times, if they should ask you. Well I have to look forward to another weary Sunday all alone and Mr. B. is thinking of going to Boston over Sunday. You take advice from our friend Mr. Gremmendohr, eh?

Well you must not listen too much to what people have to say. Because you know what I think of him. We are going to the Movies tonight. Wish it was you. Harry Thaw was said to be in town the other day, but I did not see him. There is not much news this trip, so guess I will have to postpone till Sunday.

Lots of love and kisses and hugs, from.

Herb.

May 23rd 14

Dearest darling pet,

Excuse me I should have said "little" angel pet, darling sweetheart. Enuf said. Stop your not a married man yet. Well I have been working again. The commandant is spring cleaning and I have been in framing pictures and washing glass for him. On with the dirty work. Mrs B has returned. As soon as the new bill is announced please send me a copy for the opening week. As I expect to be home Friday maybe at noon. I would be very much pleased to attend the theatre at night so be sure get two Isle seats halfway back Fri. night. I believe its customary to give the host a treat while on a visit so I guess its up to you Spike. Better purchase on Mon. morning if possible. Remember Im home to stay this time Spike. No more roaming for me. I've been feeling punk for the past couple months.

I will mail this Sat night so you ought to get it Mon. am. Its a long time between now and Friday. I go to the Movies tonight as usual. I went to bed last night. Im going to bed tomorrow night. I wrote a letter to each of the prospective candidates for operator to find out their views on accepting my position and found they were not anxious to come. One guy had a better job and the other could not board his family of three on 3 a day. The letters were intensely interestin. Ill let you read 'em. Remember me to all.

Love
Hugs
&
Kisses

Herb.

Chapter 15

Portsmouth, N.H.
May 27, 1914

Dear Grace,

We are having some hot weather for a change perhaps you have already observed it yourself. Well it was hot last night too believe me it registering 84 all night in my inclining room. Your letter received with the enclosed pieces regarding Polis new theatre, and after reading the kind of show they will produce this week, it will be of no use to purchase seats in advance and not many will be reserved. As the house is so room we will have no trouble to obtain seats any night except holidays. I thought they would run vaudeville there. If you have already purchased them, previous to receiving my telegram please call Pa. up and tell him to advise me by to that effect. Something might turn up that I did not get away Friday morning and in that case they would be wasted. Unless you take Auntie or Uncle with you. They show probably is not worth much but never-the-less they will fill the house as it is a new one., and everybody is curious to see Hartfords added attractions..

Believe me it was hot last night without very much clothing either. Mrs. B. turned in early about nine oclock. And I went down town on the car to search for amusement, but when 9:30 came I was glad to head for home and go to bed. It was so hot that I did not think I would be able to get to sleep, but when I lay down I fell to sleep immediately and did not wake up till 7 this morning. It must be nice to be single hot nights, eh? Well I dont know about that. Sunday I went for a ride all by myself to cable Road Rye N.H. and went through the cable house. They employ seven operators there and they are of all nationalities. English, Irish and sctoch, all being imported from the old country. They certainly live out in the wilderness. It is an hours ride on the car to cable road, and then you get off and walk a mile and a half and then you are there. The gink was telling me that before the cars run to the corner, he used to walk into Portsmouth, 7 miles distance, and think nothing of it. He likes Portsmouth very much, and thinks I will like it too. He said that Lockwood, who used to have my place, here, and later worked for the cable company had absolutely no use for Portsmouth.

On my way home I will stop in Boston for a short time to do some shoping and possibly may stop in Springfield on business. It depends if I feel equal to it. Business is quiet with us here when everybody is on the job. I may have to take that vacation this time when home. Cant tell yet. Best love and kisses.

Herb.

Electrical telegraph systems were suggested as early as the third quarter of the eighteenth century, and serious experiments began soon thereafter. Commercially practical systems began to take shape in both America and Europe in the late 1830s and early 1840s, as Cooke and Wheatstone in England and Morse in America introduced the electric telegraph to their respective markets.

By the early 1850s, the utility of these systems had been proven, and extensive networks of telegraph wire began to blanket Europe and the populated sections of the United States. In England and the Continent, the need to connect the various countries and parts of countries separated by water soon became apparent. Similarly in America, the need to close gaps created by the larger rivers that would not be spanned by bridges for several decades was keenly felt.

The problems were not inconsiderable. While the average citizen viewed it with awe and many considered it more magic than science, the land line telegraph as it existed in 1850 was primitive and low tech compared with what would be needed to girdle the earth underwater. The technology that would be required for a submarine cable systems was several orders of magnitude more difficult than that for a land line system.

It was as though one were to walk up to Orville and Wilbur just after they completed the first successful flight of the Wright flyer and said "OK, now that you have built an aero plane we want you to produce a complete space station and shuttle system."

The first cable of consequence was laid across the English channel from Dover to Calais in 1851, a distance of about twenty miles, and was worked using conventional telegraph instruments. The first attempt to span the Atlantic from Europe to America took place in 1857 and failed. The second of two subsequent attempts made in 1858 succeeded and for the first time in history one could send a message from Europe to America instantly.

Well, almost instantly. For some complex technical reasons, it's not possible to use standard Morse keys and sounders over the distances involved, but this wasn't fully realized by the telegraph engineers of the day and they tried anyway. Messages were exchanged but the rate was abysmally slow, probably something on the order of five words an hour. In attempts to improve this rate, voltages on the lines were increased. Eventually the cable was overloaded, burning out after less than a month of operation.

After a suspension of activities while the engineering problems and the American Civil War were resolved, attempts to bridge the Atlantic began again in 1865. The steam ship *Great Eastern* laid a line from Valencia Bay, Ireland, to within four hundred miles of Heart's Content in Newfoundland, when the cable broke.

The steam ship Great Eastern *laid the Atlantic cables in 1865-1866 and the French Atlantic cable in 1869. Carte de visite, c. 1870.*

In 1866, the *Great Eastern* tried again, successfully this time. Not only did she lay a new cable, but she was able to go back and pick up the broken end of the 1865 cable as well. A new section was spliced onto the 1865 cable and it was brought to Heart's Content; there were now two working cables. The line was opened to commercial service within a few weeks, and one could now send a telegram from New York to London for a mere one dollar per letter or five dollars per word, in 1866 dollars!

In 1869, the French could no longer tolerate the English monopoly on transatlantic communication and leased the *Great Eastern* to lay a cable from Brest, France, to Duxbury, Massachusetts, by way of St. Pierre, a French island off the south coast of Newfoundland.

In 1873 John Pender, a prominent British cable promoter, established the Direct United States Cable Company with the object of

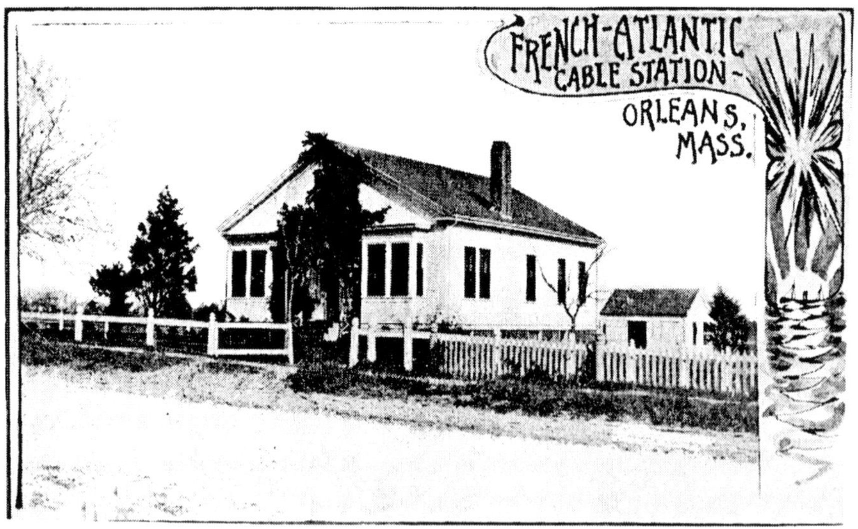

The French Atlantic cable station, Orleans, MA. The original station was at Duxbury, MA, but was moved to Orleans before 1900. The French government closed the station and locked the doors, leaving the station fully intact in the mid-1950s. It is now a museum, open summers and by appointment. Postcard view, c. 1910.

laying a cable from Ireland to the United States with no intermediate relay points. This proved to be unfeasible with the available technology; it was determined that one intermediate relay point was needed at Halifax, Nova Scotia. The continental terminations would be Ballinskelligs, Ireland, and Rye, NH. The British firm Siemans was chosen to build and lay the cable with their newly built cable ship *Faraday*.

No. 156.

MEN OF THE DAY, No. 35.

"Telegraphs."

John Pender, from a Vanity Fair *"Spy" print, c. 1875.*

The *Faraday*, at 360.38 feet long with a beam of 52.25 feet, was the second largest ship in the world at that time, only surpassed by the *Great Eastern*. She was also only the second ship to be built from scratch as a cable layer and featured a number of innovations. The *Faraday* had twin screws and a rudder at the bow as well as the stern to facilitate close maneuvering with cable, and *Faraday*'s twin stacks were mounted side by side so as to leave an unobstructed path along the centerline from bow to stern. This allowed paying out or reeling in of the cable from either end of the vessel.

During cable operations, the *Faraday* and the *Ambassador* lay at anchor in Portsmouth harbor for several days. During this period, visitors were welcomed aboard and a lady resident of New Castle described her visit in a letter to her parents thus:

New Castle N.H. 1874
Thursday Aug 13th

Dear Parents–

I have been threatning ever since we came here to write a few lines if not more to you so you might have proof positive that we had not forgotten you–

For six weeks after coming here I was so tired all the time that it was an effort for me to do what little seemed necessary to be done– I have been feeling much better a week or ten days past and feel that I am gaining fast.

We have had remarkably pleasant weather most of the time this summer now & then a shower and occasionally a rainy day– Joseph has some poorly feeling times– I trust nothing very lasting & he will come out hale and hearty after due days– He goes out on the water nearly every day he went huckle berrying yesterday three miles away on the water over to Rye–

It is just a mile across the river from here to Kittery Maine a nice little row or sail when the wind and tides are right – I often wish the children were here to see the many nice sights we have – boats or ships or steamers & some times all passing every hour in the day–

The large steamships Faraday & Ambassador were anchored right in full view of us – ten days or more Joseph went on both of them & on the Faraday twice I went aboard the Faraday only – She is a monster truly with a great amount of machinery and several trades going on her deck – a blacksmith shop with 4 men hard at work – a carpenters shop &c – They had a large hornless cow which furnishes them milk for tea & coffee – Pigs hogs–sheep–ducks–geese hens turkeys in abundance which they could slay and eat – said they had 2 bullocks when they started which they had killed & devoured –

> Such quantities of bread as we saw baking – all looked and smelled so nice (the bread I mean) & browned just such a shade all over – it requires some provisions to feed two hundred men – that was the number I think that it took to man & care for the ship – There were three tanks 30 feet deep and 20 feet across where the cable was kept – each one would hold a thousand miles of cable –
>
> And when she was anchored out here there was a large vessel went along side & discharged four hundred tons of coal on board her – It took several days – We will tell more about it when we get to Miane Frank writes us that you have got a nice barn built....
>
> J. M. & G. Thorp

Mrs. Thorp got most of it right. The cable tanks were 30 feet deep but the forward tank was 37 feet in diameter and held 400 nautical miles of cable. The midship and aft tanks were both 45 feet in diameter and held 800 nautical miles of cable each, for a total of about 2,000 nautical miles of cable. *Faraday* had a bunker capacity of 1,700 tons of coal , so Thorp's account of loading four hundred tons is entirely credible.

The *Faraday* left England on her maiden voyage in early June of 1874, bound for Tor Bay, Nova Scotia, where upon her arrival, she commenced laying the cable to Rye, NH. The cable was to be laid in several stages.

The cable ship Faraday *at anchor in Portsmouth harbor, July 1874. From a Davis Bros. stereoview.*

The shore end of the cable was shipped in another vessel, the *Ambassador*. The shore end cable was made up of deep ocean cable with the addition of several more coatings and layers of heavy armoring wire, and was laid from where the ocean started shoaling up to the beach. This was necessary to protect the cable from ships' anchors and fishing boat drags.

The Direct Cable Company cable at Jenness Beach, Rye, New Hampshire, lies exposed at low tide. Glass plate slice, c. 1900.

The cable laying had been scheduled to be completed by about mid-June but some missed rendezvous and heavy fogs caused delay. The *Faraday* was even reported to have been sunk in a collision with an iceberg; fortunately the report proved erroneous.

The June 27th issue of *Harper's Bazaar* was more optimistic but equally erroneous. *Harper's* devoted a full front page to engravings describing the landing of the new cable. In point of fact, an article in the *Portsmouth Daily Evening Times* tells us the cable was actually landed on Rye Beach on July 15, 1874. The cable terminus was initially located in the Jeremiah Locke house and was later transferred to a purpose-built building in 1875.

> According to the 1930 edition of the *Dictionary of American Biography,* the original impetus for the Direct Cable Company came from a New Hampshire native, Arthur MacArthur Eastman. Eastman was born in Gilmanton, NH, on June 8, 1810. Educated locally, he became a storekeeper in Gilmanton, then moved to Boston to engage in various trades, including the manufacture of woolen underwear and munitions. During the Civil War, Eastman made his fortune selling arms to the U.S. government and later to foreign governments. His biographer describes his involvement with the cable as follows:
>
> About 1869 Eastman planned the enterprise which was to be the great work of his life, the laying of a direct ocean cable between Europe and the United States. The difficulties in his way were

The Jeremiah Lock House as it looks today. It was the first home to the Direct Cable Company in Rye.

> great. Not only was it necessary to raise $6,500,000 in gold, but also to overcome at every step the powerful opposition of the Anglo-American, the Western Union and the French companies. Securing a charter from New Hampshire and permission from the federal government, Eastman went to Europe, where he was well and favorably known, and there, after five years of tireless effort, he obtained the necessary funds. The American end of the new cable was laid at Rye Beach, N.H., in July 1874 with elaborate ceremonies, and it was in full working order in the next year. It provided the first competition from this country with the Anglo-American Telegraph Company...

The Anglo American Telegraph Co. was owner of the original Atlantic cables and maintained extremely high tariffs, thereby barring all but the most wealthy individuals and companies from benefiting from the cable. The entire country, with the exception of stockholders in Anglo-American, seemed to be behind the Direct Cable Company in hopes that its competition would drive down prices. Grabbing a piece of this lucrative business must have been a prime motivator for the sharp businessman.

It's fairly obvious that Eastman's good connections both in the state and federal governments must have been key to getting the necessary permissions and charters for the cable, but most of the finance came from Europe. It's not certain how much stock Eastman held in the Direct Cable Company, but it almost certainly was not a majority share. It's equally certain that the majority share lay with the European backers, including John Pender who incidentally was chairman of the board of Anglo-American. The Direct U.S. Cable Company was very popular with customers and stockholders alike and was declaring larger dividends than Anglo-American. Pender took note of the damage being done to Anglo's bottom line and engineered a takeover of Direct Cable by Anglo-American in 1877, although the Direct U.S. Cable Co. continued to operate under its own name.

Herb spoke in his letter of his visit to the Direct Cable Company offices and meeting some of the operators there. The picture of the Direct Cable Company operators is from the period during which Waldron made his visit and it is likely that one or more of the operators in that photo are the very ones he met. Patrick Reib, first gentleman in the back row, was in fact the Superintendent of the Rye office. Charles Hazlett's 1915 *History of Rockingham County, New Hampshire...* has this to say about Rieb:

> Patrick W. Rieb, superintendent of the Direct United States Cable Company, Limited, at Rye Beach, N. H., has been located here for the last thirty years, having held his present position for the last two years. He was born in Dublin, Ireland, November 5, 1854, a son of John and Mary (Byrne) Rieb, the father being a watch maker.

D. U. S. Cable Station. Rye, N. H.

View of the Direct U.S. Cable Station, Rye, NH. Postcard view, c. 1915.

Direct Cable Company operators, c. 1914. Back row, left to right: Patrick Reib, Joseph Watt, two unidentified operators. Front row: John Frazier, William Frazier, one unidentified operator. Cabinet photo.

> Patrick W. Rieb was educated in Dublin by the Christian Brothers. He left school at the age of fourteen years and was then employed in a law office in Dublin for about a year. He then joined the telegraph service, was ten years with the government service in Dublin and Birmingham, during which time he edited and published the Telegraph Journal, subsequently joining the Direct Cable Company at Chester, England, being clerk in charge there for three years. He then came to America to take his present position at Rye Beach.
>
> Mr. Rieb married Annie Smith, who was born at Tipton, England, a daughter of Thomas Smith. Mr. and Mrs. Rieb have four children: William Ernest, and operator in the employ of the Cable Company; Frederick, also an operator; George, engaged in the same business; and Florence, wife of Alexander J. Yeats, an architect at Boston, Mass. Mr. Rieb is republican in politics, but votes with judgment, occasionally disregarding party line. He attends the Congregational church, which he has served as warden and treasurer.

The Direct Cable was bought by the British Government in 1920 and the Rye terminus was abandoned soon after. The Direct Cable was the last of the great pioneering cables and the last to garner front-page attention in a national paper. A large number of cables were laid after 1874 all around the world and some may have received front-page coverage in their local papers, but for the most part they would obtain only a few column inches in the business section.

There always is an exception to any rule. Several decades later one cable project did garner national attention and Newington, NH, played a role. The telephone was invented in 1876 and phone lines were being strung soon thereafter, but a true trans-continental phone call, from New York to San Francisco, was not accomplished until 1915.

It would be another forty years before a trans-Atlantic telephone call could be transmitted over a wire. To be sure, there had been trans-Atlantic telephony since the late 1920s, but it was accomplished by use of radio and that produced a number of problems. A call carried over the Atlantic by radio was subject to all the usual problems of radio, i.e. summer static, changing propagation causing loss of signal, and limited amount of radio spectrum limiting the number of calls that could be made at any one time. Technical innovation mitigated some of these problems, but one problem that couldn't be overcome was the fact that anyone with a multi band short-wave radio could listen in on your conversation. Privacy was non- existent!

In 1953, the British Post Office and the American Telephone and Telegraph Co (A.T.&T.) made an agreement to lay a trans-Atlantic telephone cable from Nova Scotia to Newfoundland to Oban, Scotland. The production of the various components, cables and repeaters, was divvied up between several British and American companies. Some of the cable for this historic telephone line was produced in Newington, New Hampshire, by the Simplex Wire and Cable Company, now a part of TYCO Telecommunications Integrated Cable Systems. The first trans-Atlantic telephone cable was completed and placed in service in 1955 and was a great success. Pacific cables were not far behind.

Unfortunately success can be fleeting. The advent of communications satellite in the 1970s and 1980s cut severely into the cable business. In the 1980s, it was estimated that only two percent of trans-oceanic traffic was being handled by cable; the rest moved over satellites. Satellites had much more capacity (called bandwidth today) than wire phone and telegraph cables. However, they were still radios and subject to some the same problems as the early radiophone circuits, not the least of which is the problem of maintenance. When a satellite gets sick, the attending physician has to make a house call, and that gets expensive.

In the last decade the pendulum has swung back. A new technology, employing fiber optics cable which transmits signals along thin glass filaments, has gained ascendancy over satellites. Fiber optic cable is relatively inexpensive to manufacture and can be laid in the same manner as the old wire cables, but has enormously greater bandwidth and reliability. Fiber optic cables now carry over eighty percent of the transcontinental traffic. Happily, Tyco Telecommunications is once again one of the premier players in this arena.

Cable Company staff picnic, c. 1880. Fom the collections of the Portsmouth Athenaeum.

Chapter 16

Portsmouth, N.H.
June 3, 14.

Dear Gracie,

I arrived O.K. but was so homesick when I got to Boston that it made me sick. The dust was terrible, and my new suit was more than a sight when I got here. My neck did certainly need a bath. I enclose the dollar with thanks to you. I did not know but I would have to come back home. It did not bother me any to leave Hfd. at first, but when I got to Boston it was all over with. Things are about the same here. My work was awaiting me and I did not get it all caught up today. I think I will not have any trouble collecting my wages for the time I was out. He received my telegram all right. Faust was out yesterday leaving Young and Hogan all alone. I hope auntie is feeling fine, and also Grandma. I'll bet Togo misses me already. Guess I will have to call it off for this afternoon, as the work is beginning to pile. Lots of love and kisses. Herb.

June 5, 1914

Dear Grace,
Yours rec'd this pm. Was glad to hear that you were glad that you loaned me the dollar and and although I did not positively need it, never the less I spent it. I knew you were lonesome when you said good bye as there were salt tears in your eyes, and you realized that it would be some weeks before you went to the theatre again. That is with me- Its six oclock and pouring hard again. The cost

of living is going up every day. She charged me board for the first day I was out, as she says that is customary, and egg sand wiches have gone fm 5 to 10 cts ea. at the yard. Dear me I dont know what I will do. Im still lonesome and Hfd. sick combined. That makes it twice as bad. But there is no use to keep reminding you of that. You have your own troubles. The kid is sick with bronchitis and they have been up two nights with her. Went to the theatre last night and witnessed a scotch Comedian but he was bad. Got on a car headed for home but stalled in front of the YMCA so got off and walked the rest of the way and beat the car home. Its stopped raining and I havent had any supper yet. Field went on his 10 mile walk this pm. This is compulsory according to the Navy regulations. This must be done once a month. Every time I write a short letter I notice you follow suit, and you offered as an excuse that you wanted to give it to the postman. Well Ive got to stop writing now because there is no news and my paper is nearly finished. Hope aunt is better. Love & kisses

Herbert.

Portsmouth, N.H.
June 10, 1914.

Dear Gracie,

Am very sorry to hear Grandma is so low, and know that it is a great strain and sorrow for you to bear, but it can not be helped and is for the best. If anything happens before I return home, please order some flowers for me. I am leaving here on Saturday or Sunday for my months leave, and probably will not return. Last night was very cold, and today it is so hot that one can hardly breath. Mrs. B. went to a wedding this morning up in the country and I will not see her again before I leave town for good.We got a circus in town today, and the Commandant and a lot of the gang left to see it. So we are alone in the office without much work to do. I am very sorry this had to happen, taking leave but I am just as well satisfied in a way. As you know things up here are not very encouraging, and the spare time becomes a bore away from all my friends. Well now I have got to make the best of it. Your wont have to write me after this Friday. Wont that be nice. I will like it better to be able to see you than write you. Auntie May has requested that you come out to see her some Sunday with me when I get back, as she is quite lonesome and alone in the world.

My best love and remembrances to the family. Also kisses.

Herbert.

June 11, 14.

Dear Gracie,

Your Wednesday letter received this morning, and was surprised to think that you did not receive my Sunday letter. You say this is your third you have written this week. I wrote you one Sunday which you should have recd, Monday am, and I also wrote you one Wednesday night which you will receive this morning (Thursday) and also this one which you will receive Friday morning. I notice you did not read carefully the contents of my Sunday letter, or else did not receive the same, as the last two letters you sent made no mention of the questions asked. Well anyway maybe you though I was joking or something else. Speaking about this place I have had enough of the Office and contents, leaving aside the impossible occurrence. I am sorry this had to happen, but at the same time I should be nearer home. I am sorry that Grandma is so bad off and that you are having so much worry and anxiety, but it had to come sooner or later. Things are going about the same here. Two more resigned the last week. Regards and best love. Herb.

Herbert Waldron never returned to Portsmouth after being laid off. He went back to Hartford and resumed his old job with Western Union where his father Arthur had become Chief Operator.

Grace and Herb were married near the end of 1916 and in 1918 or 1919 he went to work for the American District Telegraph Co. A few years later Waldron became a telegraph operator for a Hartford stock brokerage house, where he remained until he retired around 1960. Grace's sister Christina and Uncle Thomas lived out their lives with the young couple in their West Hartford home.

Grace died in December 1964 at the age of 79. Herb followed his beloved Spike four years later, at the age of 80.

Bibliography

Boyd, Captain David F., U.S.N. *Extracts from the Daily Log Book, U.S. Navy Yard, Portsmouth, New Hampshire; October 15, 1819 - December 17, 1929.* Kittery Historical & Naval Society Museum collections.

Brief Resume of the History of the Portsmouth N.H. Navy Yard December 1935. 4 page time line of significant events in the history of the shipyard; mimeograph copy. Portsmouth Naval Shipyard Museum and Visitor Center collections.

Bright, Charles, F.R.S.E. *Submarine Telegraphs: Their History, Construction, and Working.* London: Crosby Lockwood and Son, 1898.

Caldwell, Augustin. *The Rich Legacy: Memories of Hannah Tobey Farmer, Wife of Moses Gerrish Farmer.* Boston: privately printed, 1892; pg 631.

Clark, George H. *The Life of John Stone Stone.* Dan Diego, CA: Frye & Smith, Ltd., 1946.

Clarke, Arthur C. *Voice Across the Sea.* New York: Harper & Brothers, 1958.

Cummings; O.R. "Portsmouth Electric Railway." Wheaton, Illinois: *Electric Traction Quarterly*, 1966.

Cummings, O.R. "Trolleys To York Beach, The Portsmouth Dover & York Street Railway." *Bulletin No. 1*, New England Electric Railway Historical Society.

Dictionary of American Biography, Vol. 5. New York: Charles Scribner's Sons, 1930.

Dictionary of American Naval Ships. James L. Mooney, Editor. Washington, D.C.: Naval Historical Center, Dept. of the Navy, 1981.

Dorf, Phillip. *The Builder a Biography of Ezra Cornell.* New York: The Macmillan Company, 1952.

Encyclopedia Americana, The. Danbury, Connecticut: Grolier, Incorporated, 1983.

Gillette, Christine. "Seacoast's TyCom plans undersea cable network." Portsmouth, NH: www.seacoastonline.com, November 5, 2000.

Griffith, Richard and Arthur Mayer. *The Movies.* New York: Bonanza Books, 1957; pg 442.

Grosvenor, Edwin S. & Wesson, Morgan. *Alexander Graham Bell the Life and Times of the Man Who Invented the Telephone.* New York: Harry N. Abrams, 1997.

Haigh, K.R. *Cable Ships and Submarine Cable.* Washington, D.C.: United States Underseas Cable Corporation, 1968.

Hazlett, Charles A. *History of Rockingham County, New Hampshire and Representative Citizens.* Chicago: Richmond-Arnold Publishing Co., 1915; pg 1146.

Hubbard, Geoffrey. *Cooke and Wheatstone and the Invention of the Electric Telegraph.* London: Routledge & Kegan Paul, 1965.

Kloss, William. *Samuel F. B. Morse.* New York: Harry N. Abrams Inc., 1988.

Jepsen, Thomas C. *My Sisters Telegraphic: Women in the Telegraph Office 1846-1950.* Athens, Ohio: Ohio University Press, 2000.

Karney, Robyn, Editor. *Chronicle of the Cinema.* New York: DK Publishing; 1997; pg 941.

Mandell, Mel. "120,000 Leagues Under the Sea." *IEEE Spectrum.* April, 2000, Volume 37, Number 4.

Massachusetts Death Records held at New Bedford Free Public Library, New Bedford, MA.

Morris, Kenneth M. & Virginia B. *Your Guide to Understanding Investing.* New York: Lightbulb Press, 1999.

N.Y.D. Form 115 (Revised); Navy Yard record of Radio Station Bldg 103 structure, dated Nov. 1, 1919. Portsmouth Naval Shipyard Museum and Visitor Center collections.

Parsons, Langdon B. *History of the Town of Rye...* Concord, NH: 1905.

Prime, Samuel Irenaeus. *The Life of Samuel F. B. Morse, LL.D., Inventor of the Electro-Magnetic Recording Telegraph.* New York: Appleton and Company, 1875.

Rayne; Mrs. M. L. *What Can a Woman Do; or Her Position in the Business and Literary World.* Detroit, Michigan: F. B. Dickerson & Co., 1884.

Reid, James D. *The Telegraph in America: Its Founders, Promoters and Noted Men.* New York: 1879.

Russell, W. H. *The Atlantic Telegraph.* London: Day & Son Limited, 1865.

Scott, John F. *Brudder Bones' Book of Stump Speeches, and Burlesques Orations…* New York: Dick & Fitzgerald, Publishers, 1868; pg 188.

Sobel, Bernard. *A Pictorial History of Burlesque.* New York: Bonanza Books, 1956; pg 194.

Tarrant, D.R. *Atlantic Sentinel: Newfoundland's Role in Transatlantic Cable Communications.* St. Johns, Newfoundland: Flanker Press, Ltd., 1999.

Varrell, William M. *Rye on the Rocks: A Tale of a Town.* Boston, MA: The Marcus Press, 1962.

Manuscripts

G. Thorp letter to parents. New Castle, NH: 2 pg ALS, Aug 13, 1874. Author's collection.

Herbert D. Waldron letters to Grace I. Glen. Portsmouth, NH: 34 ALS and TLS, Mar thru June 1914. Portsmouth Athenaeum collections.

Periodicals

Geer's City Directory. Hartford, CT. 1880-1970.

City Directory. New Bedford, MA. 1850-1910.

Harper's Weekly. New York, NY. 1857-1914.

Hartford Courant. Hartford, CT. 1960-1970.

The Daily Evening Times. Portsmouth, NH.

Portsmouth Herald. Portsmouth, NH.

Portsmouth Journal. Portsmouth, NH.

Portsmouth City Directories. Boston: W. A. Greenough & Co., compilers. 1840-1915.

United States Census. 1860-1930.

Web Sites

www.ancestory.com - This is a subscription site but may be available for free at your public library.)

www.familysearch.org - The Church of Latter Day Saints (Mormons) genealogy site contains genealogical information on people of all faiths throughout the United States and many foreign countries.

www.google.com - Search engine on the World Wide Web.

www.uscg.mil/hq/g-cp/history/weblightships.gov - U.S. Coast Guard history site; including information on the Coast Guard and the U.S. Lighthouse Service.

ABOUT THE AUTHOR

Bill Holly was born in Middletown, New York, in 1939. His family moved to Colonie, New York, near Albany, soon thereafter. Graduating from Colonie Central High School in 1957, he joined the U.S. Coast Guard the same year. After graduation from radio school in Groton, Connecticut, Bill saw duty on the Great Lakes, and in Alaska, Florida, Puerto Rico, and New England, as well. While in Puerto Rico, he married his high school sweetheart, Fran. In June 1979, after 22 years service and over fifteen years at sea, Bill retired from the Coast Guard as a Chief Radioman and began his second career as a master electrician at Portsmouth Regional Hospital, Portsmouth, New Hampshire. the following month. In April, 2001, after 21 years, Bill retired from the hospital.

Bill has been an amateur radio operator since 1964 and began collecting antique radio and telegraph instruments and studying their history in the early 1970s. He has written articles for *QST*, *CQ*, *The Old Timers Bulletin*, and *Classic Toy Trains* magazines, as well as a monograph for the *AWA* [Antique Wireless Association] *Review*. This is Bill's second book, the first being a definitive history of the Vibroplex Company.

Bill and Fran live in Kittery Pt., Maine, and are currently employed as full time staff for two cats. Their two grown daughters, Robin and Kate, have long since flown the nest, but live near by in almost tax-free New Hampshire.